The Little Book of the Earth, Sun, and Moon

by Dan Browning

Vesta Imprints
San Francisco

The Little Book of the Earth, Sun, and Moon

Published by Vesta Imprints

2nd Printing with Corrections

ISBN: 978-1-7347962-1-6

Table of Contents

Chapter 1: Inner Space

What if everything is wrong? Is science too conservative? Is there some value in what can be called *reckless science*? Reckless science doesn't wreck the universe; it looks outside the box and questions all the things that we were taught as children; and all the things we grew up believing.

One way to reverse one's outlook on the world and the universe is to reverse everything that's "known". Let's begin.

We think that the Earth is part of a solar system. Ten or so planets orbit the Sun, and the Sun, a star, orbits the galactic center every 225 million years. The Sun is supposed to be on the outer edge of a spiral arm of the galaxy, far away from the galactic center.

Let's reverse that and put the Sun at the exact center of the entire galaxy. The Sun doesn't orbit the center of the galaxy; rather, everything in the galaxy orbits the Sun every 225 million years.

We think that the Earth orbits the Sun. Copernicus was responsible for that revolutionary thought (no pun intended) and we live in the Copernican Age. Before the mid-17th century our primitive minds could only conceive of the Earth as being at the center of the universe. In 1633 the Roman Catholic Church condemned Galileo for teaching that the Earth goes around the Sun. Galileo had to renounce Copernicus. Only in 1992 did the Church accept the Copernican system and pardon Galileo. Well, forget all that. The Greeks believed the Earth was at the center of the universe, surrounded by spheres, including a sphere of the Moon, a sphere of the Sun, two spheres for the known planets, Mars and Jupiter, and a sphere of the stars.

Let's reverse the Copernican revolution and place the Earth inside of the Sun. Since the Sun is, in our reversed universe, at the center of the galaxy, one can say that the Earth really is at the center of the universe. The Earth doesn't orbit the Sun every 365 days: the Sun orbits the Earth every 365 days. Such reversals can be thought of as "relativity shifts". Einstein taught us that everything is relative.

We think that life exists on millions of other planets in the galaxy. Planets that orbit at a particular distance from their stars are the most likely to harbor life, because with just the right amount of light and heat those planets have life-giving conditions.

Let's reverse that and say that life only exists on Earth. Why? Because life can only exist when a planet enjoys the protection of being both inside of the Sun, and of being at the slow-moving center of the galaxy. Planets that orbit stars that reside outside of such a privileged place are just too unstable to support life. Life only exists on Earth.

We think that, at the center of each galaxy, the galaxy is powered (its rotation and magnetic field) by a massive black hole. Black holes supposedly can contain no life because anything that falls into a black hole is ripped apart by the tidal forces of gravity.

Let's reverse that and posit that life can indeed exist inside a black hole. Not only can life exist, but life can *only* exist within a black hole. Not only can life only exist within a black hole, but life can only exist within a black hole at the center of the galaxy. All other black holes that may orbit exterior to the central black hole are too small to support the many interior horizons that surround the water reservoir at the center. Those horizons provide protection. Everything that orbits outside of the Earth is in fact first lensed through these multitudinous horizons that extend far out into space.

The universe is 13.5 billion years old and the Earth itself is 4.5 billion years old. 3.8 billion years ago plate tectonics got started on the Earth and Moon. 2.3 billion years ago the partial pressure of oxygen at the Earth's surface increased, leading to the Great Oxygenation Event and the development of complex lifeforms on Earth. Life on Earth is certainly not older than 4.1 billion years, when bacteria and algae existed in the oxygen-poor waters.

We think that, in our solar system, water condenses out in the vicinity of the Earth, and silicates condense out in the vicinity of the Sun. The context here is the condensing out of elements, both in the current timeframe, and from the cloud of gas and plasma that the solar system was originally formed from. Methane, on the other hand, condenses out farther away, near Saturn and composes the methane oceans and rivers of its moon Titan. Jupiter only has 10 to 20 percent the abundance of water that the Sun does.

Let's reverse things and say that the greatest abundance of water exists at the galactic center. Without water, there would be no galactic center, and no Earth. More, a black hole can't exist *without* water.

Scientists know that a star goes through several stages in its evolution. Stars burn both hydrogen and helium at their cores when they initially expand. But because of their enormous masses, they are capable of igniting additional nuclear reactions. In a high mass star such as the one at the center of the galaxy, the tremendous weight of overlaying stellar material forces central temperatures to 700 million degrees, and carbon burning begins. Still later, when temperatures have risen to one billion degrees, oxygen burning is ignited. In each case, the reaction proceeds at the star's center until all the fuel is used up. The reaction then momentarily (on a cosmic scale of time) shuts off and the star's core contracts under the influence of gravity.

Temperatures immediately above the contracting core become so high that a new round of nuclear burning is ignited in a thin shell around the core. The ash of oxygen burning is silicon. As a thin shell of burning oxygen moves outward, away from the star's center, an ample supply of silicon is left behind. And when further compression of the core has forced the central temperature to about 3 billion degrees, silicon burning is ignited. Iron is the ash of silicon burning. But iron does not burn, no matter how hot the star's core has become. Near the end of its life, therefore, a massive star has an inert, iron-rich core surrounded by several thin shells in which thermonuclear reactions are still occuring. These nuclear burning shells are crowded close to the star's core.

We think that the rising ocean levels are due to the melting of the ice caps because of global warming.

Let's reverse that, and say that when ocean water is heated, it expands. There are lanes of expansion that circle the entire planet in two great circles: one above the equator and aligned with the plane of the Sun's orbit; and one below the equator and similarly aligned. Because the Earth is tilted relative to the plane of the Sun, the bands are also tilted. These bands are where the ocean has risen the fastest. If the internal heat source of the Earth is also the internal heat source of the Sun and the Galaxy, that means that whatever's happening is happening at a grand scale.

If the oceans are the result of hydrogen burning and act as a cooling system for the Sun, they should be prized and protected. Scientists at Northwestern University found that pressurized hydrogen is stored in a rock layer deep beneath the Earth's surface, in a kind of rock called ringwoodite. Plate tectonics is much more than the moving of surface-crustal plates; it also renews the ocean's water supply by injecting water trapped in mantle material that upwells with the tectonic motion. The scientists call this the "whole-Earth water cycle".

We think the Moon is the coalesced matter from an asteroid hit on the Earth, and thus made from the same elements as the Earth. This is called the *catastrophic theory* of the origin of the Moon.

Let's reverse that, and say that even though the elements of the Earth and the Moon are similar, the Moon's highlands are high in silicates and *refractory*, meaning they were formed under high temperature and high pressure. Let's return to the Earth and Moon's evolutionary history. As mentioned, plate tectonics got started at the same time on the Earth and Moon. Radiocarbon dating of rare-earth elements in the rocks of the Earth and Moon show that both planetary bodies started to cool and solidify at exactly the same time: about 4.51 billion years ago.

Both bodies developed proto-atmospheres almost as soon as cooling began. 3.8 billion years ago the proto-atmospheres of the Earth and Moon were thick with chemical compounds that rained down on the respective planetary surfaces. The proto-atmospheres contained hydrogen and oxygen, and therefore water. The planetary scientist Gerard Kuiper estimated that over a period of 100 million years, the Moon and Earth were protected from outside UV radiation due to these thick atmospheres, but after that, as the interface between the Earth and Moon became superheated because of silicon-burning in the outer shell, the proto-atmospheres evaporated, and the Earth and Moon formed secondary atmospheres.

The idea of the Moon orbiting the Earth in a circular orbit, in the shape of a sphere, though true on one level, may not be the only way to see things. Stellar burning regimes are better understood as operating on a system of concentric spheres. At one level the Sun's chromosphere floods outer space with light and heat, and the Earth is at just the right distance to be heated by that radiation. But at another level the Earth may reside inside of the Sun and receive the Sun's benefits in the exact same way.

The Earth's core can be seen as the Moon. In a concentric sense, the Earth and Moon are connected by the Earth's mantle. But whereas the Earth's surface retained its water, the water of the lunar shell was largely evaporated or possibly downwelled (gravitationally stolen by the Earth) over the course of 800 million years starting 3.8 billion years ago. The remaining oxygen between the Earth and Moon was converted to heavy isotopes that burned out, or was pulled to Earth by gravity and the Earth's developing magnetic field. Meanwhile, the silicon in the lunar shell became volatile due to the presence of radioactive elements that now reached the Moon due to the loss of its secondary atmosphere. Also during this period volatiles upwelled from the Moon's mantle, which is the closest the Moon came to plate tectonics.

Silicon burning came to an end some 3 billion years ago. The Moon's surface cooled, but close beneath the surface the Moon remains hot. The Earth's mantle and iron core remained plastic and hot, and since iron can become hot but doesn't burn, the Earth's iron core remains hot to this day. Fundamentally the Earth's core is the core of the Sun itself because Earth is at the center of the Sun. 2.3 billion years ago an oxygenation event occurred on Earth, with the star's original and now recycled oxygen and carbon becoming the key ingredients for life on the Earth's surface. Meanwhile, the Moon started a new phase of its evolution….

We think that the craters on the Moon were formed by impacting meteors over time. Craters were first called *circular mountains* by astronomers in the 19th century. By the 1920s, German scientists started using the term *ringwall*. But there has been an ongoing debate about their origin. Many support the impact theory, but others support the theory of volcanic origin.

Let's go even further and give craters a new name: *gamma flux-tube radiation resonance rings*. The idea is that craters are the resonant artifacts of incredibly powerful radiation processes going on in the Sun. The movement of crustal material from the interior of the crater to the ringwalls (and beyond) hasn't happened instantaneously, but over billions of years of evolution. The effects of radiation resonance have much in common with volcanism, but at a very slow rate of change. The other principal dynamic is surface erosion due to electric charges. Over billions of years craters have been excavated to greater or lesser extents depending on the strength of the flux. As the potential of a major crater fades away after billions of years of activity, the flux disperses into smaller emanations and thus produces smaller, fresher craters.

A full catalogue of the Moon's volcanic features will be covered in Chapter 2, but suffice to say that the Moon's surface contains calderas, primary and secondary faults, fault-zone depressions, rift spreading, volcanic corridors, volcanic domes, and fumaroles. The older craters are filled with dust that has eroded away over hundreds of millions of years. In a thick layer of dust, the force between the particles of a given size will increase linearly with depth from the surface and hence so will the area of true contact. Under such conditions, the Moon's natural radioactivity will affect temperature.

The temperature of the Moon's surface starts to get very hot very quickly with increasing depth. At a 1 meter depth, the temperature is estimated to be 240 degrees Centigrade, or 432 degrees Fahrenheit. At 25 meters depth, the temperature will reach the softening point of basalt. This means that dust cannot lie indefinitely over large areas at depths of more than about 25 meters because such lunar material will show a temperature rise in the region of 8 degrees Centigrade per million years. Even if the craters were not formed by volcanic action, the mere filling of the larger ones with dust would provide, in time, for volcanic activity at their centers.

Infrared images of the Moon show that the hottest part of a crater is its center, where erosion has creating perching of hemispherical and pyramidal fragments some 5000 feet in altitude. These perches or pedestals are more resistant to erosion than the material around them. The pedestals at the crater centers have a higher surface roughness on a scale of centimeters than the surrounding material, and are therefore hotter. As well, the pedestals absorb more heat because they carry more of an electrostatic charge. Electrostatic repulsion sometimes causes a *dust halo* to form around these features. What's more, even though the distribution of craters appears totally random, the hot spot distribution is *non-random* and conforms to an underlying pattern of basaltic corridors. The hottest craters overall are the newest ones, their temperatures a direct consequence of the hot sub-surface dynamics that caused them in the first place.

The Apollo scientists were worried about the astronauts sinking into the dust, but their worries were unfounded. The topmost surface of the Moon is *baked*. Over 1 billion years the heavy atoms on the lunar surface have been significantly enriched – that is, they've become more radioactive and hotter – and loose dust particles have been fused by sputtered atoms produced by ionizing radiation. Radiation resonance has therefore converted the top few inches of the Moon's surface into a porous but rather firm crust. The surface was hard enough to support the astronauts, the Moon buggy, and the lunar lander at one-sixth Earth gravity. That said, the astronauts still had to deal with a thin layer of loose dust, as photos of their dust-smeared hot sweaty faces after returning from an EVA (extra-vehicular activity) make plain.

We think of the Sun and Moon as fixed bodies of a fixed size. The Sun is 864,000 miles in diameter and the Moon is 2176.

Let's reverse that. The science-fiction author Arthur C. Clarke postulated that the Sun could be ten times nearer and ten times smaller and we'd never know it. By the same token the Moon could be ten times farther away and ten times larger. By a simple ratio, the assumed distance from the Earth to the Moon, 238,000 miles, when expanded to the assumed distance to the Sun, 92.8 million miles, results in a diameter of 2176 x 92,800,000 / 238,000 = 848,457 miles – nearly the same diameter as the Sun. The Moon and the Sun are said to have the same *apparent diameter*. More precisely, 238,000 miles as a radial distance is the same as a diameter of 476,000 miles, and this in turn can be thought of as one measure (but not the only one) of the lunar shell inside of the Sun.

Even more precisely, the true nature of the event horizons around the Earth is that they dictate all things "Sun". The old way of doing things involves distances and velocities that correspond to the perceived rotation of the Earth around the Sun. The new way of doing things uses Einstein's relativity theory. Einstein's theory says there are three event horizons around the Earth. Instead of computing *circles* for orbits, you use a *sheet*. There are three sheets for the three event-horizon distances. Each sheet is the surface area of a geometry such as a spinning top. The process for deriving orbits is similar to chart-making for aeronautics, or for ship navigation, in the sense of navigating by "orbital lines" or vectors.

Thoughts

We look out towards the stars and count our achievements by the number of colliding black holes we can observe at the laser-interferometry observatories around the world. These black holes are billions of light years distant. But new laser-interferometer projects will start measuring gravity waves in space, and look inwards rather than out to the edges of the universe. What might the interior of a star look like? What's inside of the Sun? Gravity waves penetrate the layers of a star instantly. A gravity-wave telescope in space could tell us what's inside of the Sun – give us the whole picture.

The twenty-first century has the potential to be the Century of Truth, but it's all up to the scientists who run the programs around the world. I give them the benefit of the doubt, along with our new generation of educated youth who will be on the front lines of new discovery.

Chapter 2: Precious Water

Finding Truth in Diamonds

In the 18th century Henry Cavendish discovered that water forms from burnt hydrogen. You can think of water as the ash of burnt hydrogen. Water itself doesn't burn because it's already given off as much energy as possible from reacting hydrogen with oxygen. When you think of the Sun as burning hydrogen, the Earth can be thought of as an interior "water reservoir". But just how extensive is that reservoir?

On the International Space Station oxygen is produced from electrolysis. The electrolysis of water produces hydrogen and oxygen. The hydrogen is then reacted with carbon dioxide

$$3H_2 + CO_2 \rightarrow H_20 + CH_4 + O$$

to form water and methane plus oxygen. The methane is exhausted into space, while the water is fed back into the system. However, two hydrogen atoms are lost for every oxygen molecule that's created, so this means the supply of water must be replenished to make more oxygen. This is not a closed-loop system. Still, it shows that carbon dioxide can be combined with hydrogen to produce methane, which is a hydrogen product, and water, another hydrogen product.

A large concentration of oxygen and hydrogen in the Earth's mantle, and carbon dioxide from the early formation of the planet, produces water and methane under pressure. Diamonds crystallize from the carbon derived from carbon dioxide or methane gas at high temperatures and high pressures. The graphite/diamond stability boundary is a sloping line that represents a tradeoff between temperature and pressure. Just enough pressure and a diamond will stabilize; too low a temperature and you get graphite.

Two types of diamond are prevalent, and both are older than the kimberlite lava that surrounds them. Diamonds are dated from their micro-inclusions. *Harzburgitic* diamonds usually reside in continental cratons – very old platforms of rock dating back to the Earth's Archean period. These diamonds are 3 billion years old. In the Archean period there was a unique process accompanying crustal formation: a combination of melt extraction, fluid interaction, and diamond crystallization that left a relatively cool, rigid keel deep beneath whatever continental plates existed back then. Think of a craton as a ship floating on magma.

The other type of diamond is *eclogitic*, which is associated with magmas affected by oceanic and continental plate subduction. These diamonds are 3 billion years old or younger and contain carbon isotopes associated with organic matter (in part) carried into the mantle by the descending slabs of plate tectonics.

Only in the 1990s did scientists find a correlation between diamonds and a deep reservoir of water in the Earth's mantle. Deep water and diamonds are intimately related. The discovery in 1999 of a strongly hydrous ringwoodite in a diamond from Sao Luiz, Brazil was the first significant evidence of water in the Earth's mantle. Since ringwoodite constitutes 60 percent or more of the lower part of the transition zone, between 525 and 660 kilometers in depth, there's likely a huge amount of water in this region, comparable or greater in mass to all of the Earth's oceans.

The major source of this water reservoir is primordial water stored early in the Earth's evolution, at a time when the Earth's atmosphere was rich in carbon dioxide.

Water in olivine and garnet inclusions trapped in diamonds are just two of the indicators for water in the mantle. These indicator minerals come from the transition zone between the deep mantle and the upper mantle. Indicators consist of cratonic lithosphere peridotites, which are *differentiated minerals*. This means that subduction was already taking place in the Earth's early history when these minerals were formed. Peridotites include green chromium diopside, garnet peridotites like red chrome pyropes, and olivine, which is the main component of peridotites.

In 2014 an analysis was performed on a microscopic ringwoodite inclusion about 40 micrometers in its longest dimension, found in a so-called “ultradeep” diamond from the Juína district of Mato Grosso, Brazil. Ultradeep diamonds form within the transition zone, and include minerals formed at the same depth as the surrounding diamond, created either at the same time or before the diamond grew around it.

The tiny, greenish inclusion allowed researchers to directly measure the water content of ringwoodite from Earth’s mantle transition zone. Through a combination of X-ray diffraction, Raman spectroscopy, and infrared spectroscopy the team showed that their small sample contained about 1.4 percent water by weight. From this, they calculated that the bulk water content of the mantle is about 1 percent by weight: an enormous volume of water.

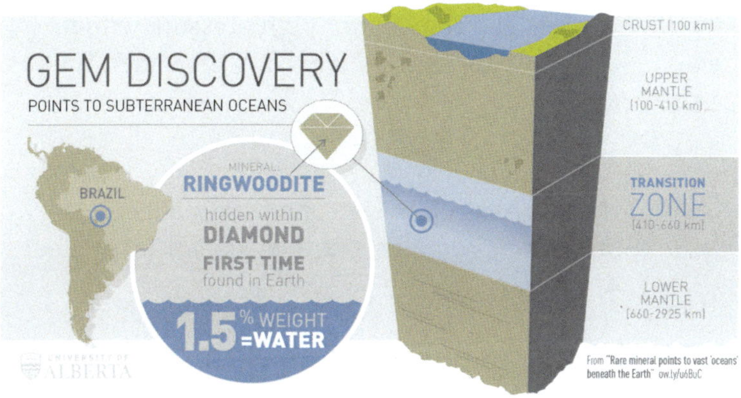

Readers may remember Kurt Vonnegut’s Ice 9 from “Cat’s Cradle” – but there’s a non-fictional type of ice called ice VII. This is a cubic crystalline form of ice formed from liquid water above 3 GPa (30,000 atmospheres) of pressure. The transition zone where diamonds are formed in the Earth’s mantle, at 660 kilometers depth, has a pressure of 23 GPa, significantly greater than the minimum for ice VII, but it turns out that ice VII inclusions have also been found in diamonds.

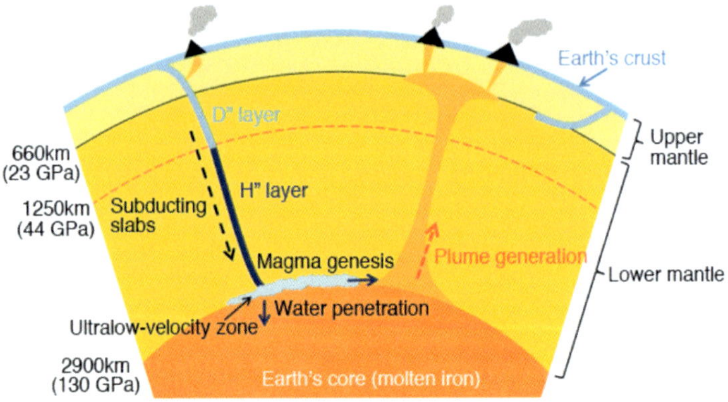

After ice VII was found in diamonds, the International Mineralogical Association classified ice VII as a distinct mineral. The ice VII was presumably formed when water trapped inside the diamonds retained the high pressure of the deep mantle due to the strength and rigidity of the diamond lattice, but cooled down to surface temperatures, producing the required environment of high pressure without high temperature. Ice VII is also thought to exist in the underground ocean of Jupiter's moon Europa.

Nobody is sure yet exactly how water forms not only at the mantle transition zone, but in the deep mantle right up to the boundary of the Earth's core. But apparently it exists in large quantities at great depths.

The idea of water being burned hydrogen is intriguing when considering the stellar model of black-hole formation, in which a series of shells burn out as the star steadily collapses into a black hole. The Sun fuses hydrogen atoms to make helium in its core. It fuses about 600 million tons of hydrogen every second, yielding 596 million tons of helium. Not surprisingly, helium content in diamonds has also been assessed and verified.

This doesn't mean that the Earth is somehow a Sun that has been turned inside out. And yet water as burned hydrogen combined with the image of the Earth as a water planet does paint a startling picture. Plate tectonics got started on Earth some 3.8 billion years ago – 800 million years before the oldest diamonds were formed. The early Earth's atmosphere, which contained chemicals which would have been subducted along with the first slow crustal masses, was rich in water and carbon dioxide. Scientists are trying to figure out how the Earth's iron core might have condensed out from precursor materials, but water in the deep earth does seem to have played a role….

The History of Earth

The Earth and the Moon formed at exactly the same time, about 4.52 billion years ago. Earth was relatively cool between 4.5 and 4 billion years ago, and no magmatic rocks have been found that are that age. Earth didn't even have a core at that time but probably started as a cool gravity-bound shell that first expanded, and then contracted. Earth also had no water at that time, but the Earth and Moon, in fits and starts, developed proto-atmospheres almost as soon as cooling began.

After expanding during the cold phase for 100 million years or so, the Earth began to contract under increasing temperatures. Over the course of 10 to 20 million years the Earth gradually became heated. Part of the heat energy came from the tidal energy of the Earth's binary interaction with the proto-Moon. When the Earth started to heat up, a mantle started to form around the then-Earth's equator, related to the path of the orbit of the Moon. Today the Earth is still affected by lunar tides, but back then things were a little more drastic.

Four billion years ago the Sun was 30 percent less luminant than today and the lunar disk was undifferentiated. There were no lunar mares. The space between the Earth and Moon was full of dust and chunks of rock. Thinking concentrically, this points to a single gravitational system of horizons. The dust and larger planetessimals were trapped by the increasing gravity of the Earth, and as the debris fell to the Earth's surface it caused heating and also introduced carbon dioxide and water from the stellar cloud. During this hot phase the

Earth formed a water and carbon dioxide atmosphere.

According to one theory, the Earth rotated once in 3 hours, so that the Sun rose on the Eastern horizon every 3 hours. The year lasted 1500 of these short days or 187 24-hour days. The Moon rapidly orbited the Earth in the same direction as the fast Sun. The Moon went through all its phases, from new to full to new again, in just 6 to 8 hours. The apparent size of the Moon was 300 to 350 times larger than today – it was immense! Its proximity alone heated the Earth's surface due to tidal interaction. At the same time, the Moon was rapidly receding from the Earth, and the lunar surface began to show fractures due to internal stresses and tidal swells.

The Moon's apparent optical gargantuan size can be explained as well by primordial event horizons very close to the Earth's surface, or possibly the exact reverse: very far from the Earth and thus providing greater magnification of the Moon, which, fracturing and heating, was actually the interior surface of the Earth, what we would now recognize as the Earth's molten core. This would explain the corollary theory of the Moon and Earth having the same angular velocities at the time, and the theory of the Moon appearing to orbit the Earth at the Roche limit – the boundary inside of which an orbit becomes unstable.

Today the Moon's distance from Earth is 384.4 thousand kilometers and the lunar tide on Earth has a swell of 46 centimeters. Four billion years ago, the swell of the lunar tide on Earth was 7 meters! The black-hole event-horizon theory doesn't explain well the presence of lunar minerals on the Earth's surface that would pre-date the 4-billion-year-old cutoff date for magmatic rocks. However one must remember that the Earth and Moon essentially formed from the same primordial cloud. Only their heating regimes differed. It may be that very old zircons found on Earth in Australia, dated to 4.3 billion years, had the same origin as zircons on the Moon, of the same age.

The Proto-Moon, as opposed to the Earth, was almost entirely molten soon after its formation, and went through a rapid differentiation. As well, the Moon seems to have had a molten core, unlike the cool

Earth of the time. But one might also contend that the Earth's core remained submerged, and so measuring the Earth's radiogenic age through available terrestrial surface rocks may not prove anything. But neither will differentiated lunar rocks of very old origin prove anything either, in terms of the Earth, even with a black-hole formation model. The simplified model is that the Earth and Moon shared and continue to share the same mantle interior. This model may seem like nonsense since the Earth and Moon are separate spheres; but they can be separate and adnate simultaneously in a four-dimensional spatial universe.

Gravity acceleration as well as a pressure increase within the Earth finally led to the "release" of a hot core starting 4 billion years ago. 3.8 billion years ago the Earth was plastic enough for plate tectonics to get started. At the same time the proto-atmospheres of the Earth and Moon were thick with chemical compounds that rained down on the respective planetary surfaces. The proto-atmospheres contained hydrogen and oxygen, and therefore water. The core formed because of drastic gravitational instability in the Earth, which led to accentuated or zonal, as opposed to global, differentiation within the Earth. At the same time, large-scale melting of the Moon's mare basalts, at depths of 200 to 500 kilometers, started, and lasted for 600 million years.

On Earth, a dense ring of molten iron, positioned at the Earth's then-equator, circled the Earth's primordial heart. At first the Earth was noticeably more heated in lower latitudes than in the upper, and for this reason the first asthenosphere and the associated zone of iron separation emerged in the upper mantle exactly under the equatorial belt of the then-Earth. In the Middle Archean, the heating of the core expanded to mid-latitudes and also a middle mantle was created. At the end of the Archean, 2.6 billion years ago, the differentiation zone covered most of the Earth.

One can also think of this as the main dynamic behind the magnetic field of the Earth transitioning from a quadrupolar field to a dipole. The Moon retained a remnant of the primordial quadrupolar field, and the Sun retains a strong quadrupolar field to this day. Early core formation was accompanied by a huge release of energy, calculated to be on the order of 5 x 10^{37} erg, which went into the heating of the mantle and the matter at the core. It's probably no coincidence that the release of the Earth's core was paralleled by the Sun heating up and becoming brighter.

Plate tectonics involves some kind of convection-structure whereby mantle flow descends towards the core. Between 3.8 and 2.6 billion years ago, immediately above this mantle flow, the first supercontinent formed. Russian scientists call this supercontinent Pangea O. This supercontinent was also the birthplace for the oldest diamonds, some 3 billion years old. 2.3 billion years ago, during the Late Archean - Early Proterozoic transition, the partial pressure of oxygen at the Earth's surface increased, possibly due to profound changes in the nitrogen biocycle. This is called the Great Oxygenation Event, and it shows up in the isotopic compositions of rocks examined from that period, including the carbon isotopes of eclogitic diamonds.

Nitrogen hasn't been mentioned in this history, but it was probably present from the beginning. A mixture of molecular hydrogen (H2) and molecular nitrogen (N2) in a ratio of 1:8, along with trace elements, which scientists now believe was the composition of the Earth's early atmosphere, will produce collisions that energize the molecules and prompt them to absorb infrared radiation. This is how the Earth was able to sustain liquid oceans in its early history, when the Sun was believed to be 25 percent dimmer than it is today. These molecular collisions raised the planet's surface temperature by as much as 60 degrees Fahrenheit by keeping the hydrocycle liquid.

The Moon is Hot

The Moon is often depicted as a cold, forlorn, and remote place. However the Moon is far from cold. It's a veritable showcase of volcanism. If the mass of the Earth is wrong and if the Moon is the interior surface of the Earth, then the Moon would have to be hot. Lunar science in the early 1970s, based in part on the observations of the Apollo moon missions, shifted the consensus from “cold and remote” to “close and hot”, and proved just how hot the Moon really is.

Lunar rock samples show similarities in chemical composition to their terrestrial counterparts. There are, however, enough differences in the crystal structure to strongly suggest that the Moon had a similar but not identical origin to that of the Earth. That is to say, the Moon appears not to have been formed from the Earth.

Though not all craters appear to be volcanic in origin, many of the larger craters resemble terrestrial calderas: volcanoes whose central portion has collapsed into the void left by ejected material. High resolution photos – for example, the Lunar Orbiter V photo of the floor of the crater Tycho – disclose a texture strongly suggesting lava flow (that is, a ropy and fluid look to the solidified masses). Though calderas on Earth are much smaller than lunar craters, the two share upwelling and thus lava flow and subsequent hardening as a characteristic.

The rays emanating from craters such as Tycho show no shadow reliefs when the Sun is at an angle to the Moon’s near side. During the full moon when the Sun is directly overhead the rays appear lightest, and the maria darkest. A safe assumption is that the maria and rays are principally coloration phenomena. Basalt is a hard, dense, dark form of volcanic rock, so that the basalt of the marias is naturally darker than the surrounding surface rocks. The color of the rays yields no clues as to how they were formed. Many rays extend remarkable distances over the Moon’s surface. The rays associated with Tycho extend more than 2100 kilometers from the crater.

It’s thought that the alps around Mare Imbrium were formed after the floor of the mare was laid down. Their origin seems to be volcanic. The so-called Alpine Valley in this landscape slashes through the alps in a nearly straight line. It has a length of about 185 kilometers and width of 21 kilometers. Photos by the Lunar Orbiter V revealed the Alpine Valley to be a fault-zone depression, not an ejecta or impact scar. It may even be a plate-boundary because of the minor spreading to either side of the fault line, similar to the spreading seen in the South African Great Rift Valley.

The so-called rilles are also fault structures. Details of some of the rilles show strings of small craters along the course of the rille, similar to volcanic corridors on Earth. Photos of the Marius Hills near the center of the vast Oceanus Procellarum, taken by Lunar Orbiter III, show ridges and domes of volcanic material stretching for hundreds of kilometers.

Orange-reddish soil on the Moon is characteristic of volcanic fumaroles. On Earth, fumaroles are deposits from the last gasps of dying volcanism wherein outpourings of gasses and water vapor oxidize the surrounding soil to the characteristic reddish color. Some of this reddish soil was found at the Taurus-Littrow site of the Apollo 17 landing.

The first radioactive dating of lunar rock samples found the age of one rock to be 4.6 billion years. More recent dating shows the oldest lunar rock samples to be the same age as the Earth: 4.56 billion years old. Basaltic material from the Sea of Tranquility was dated at 3.7 billion years old – conclusive evidence that the basaltic material in some of the maria at least was extruded, and crystallized out, about a billion years after the Moon was consolidated. This is further confirmation of the 3.8 billion-year-old critical point mentioned above.

As far as the Moon's origin, even in 1974 the "tidal theory" of the Moon being torn loose from the Earth was in trouble. The tidal theory has serious defects. Any gravitational force that could succeed in wresting the moon material from the Earth would have to greatly exceed the Earth's gravitational hold upon it. Probably only a passing star could accomplish this feat. If this were the case, how did the organization of the solar system survive the intruder's presence?

Furthermore, any such gravitational larceny would deprive the Earth of all of its water and most of its atmosphere before the intruder succeeded in capturing a substantial mass of the Earth's rocky mantle. The question of how the Earth acquired both a new ocean system and a new atmosphere poses formidable difficulties for the acceptance of this theory. A halfway theory posits that the proto-Moon simply fell apart and then, somehow, the Moon as we know it formed afterwards: again untenable because of the magical appearance of a new gravitational center.

Another objection is raised by the fact that analysis of earthquake waves discloses no scar tissue (which most assuredly would remain from such a violent event) in the Earth's mantle. Such scar tissue, however, would not necessarily be present if the wrench occured either before the Earth was consolidated or before the surface was molded by subsequent heating and melting of the mantle.

Finally, how did this "liberated" material break the gravitational bonds with the intruder and remain in the immediate vicinity of the Earth? It would seem an incredible coincidence that the passing body could tear the matter loose from the Earth, yet be traveling so fast that it got out of gravitational range before the moon matter could catch it. Today scientists agree that the Earth and Moon consolidated at the same time, and thus the formation was evolutionary rather than cataclysmic.

The entire Earth-Moon-Sun system was formed from an accretion of matter contained in the vast interstellar nebula which preceded all of the members of the solar system. This "accretion theory" is more plausible than the tidal theory, but here the principal objection rests in the fact that the Earth's larger mass should have captured the material first, instead of permitting a new gravitational center to form nearby. Today one can say that the Earth did indeed capture this material, that diamonds and zircons can be found on the Earth and Moon alike, and that the Earth and Moon formed in close proximity as two parts of a whole.

Transitions

Speaking of transition zones, we are in a transition of science that requires new ways of thinking about things we normally take for granted. We must be able to see things in seemingly conflicting ways to arrive at a full understanding – an understanding which is currently out of reach. To fully understand the Earth's place in the galaxy will require a mixture of a rudimentary math knowledge, geophysics knowledge, planetary science, observations from diamonds and other rocks, observations from satellites and rovers, and general intuition and intelligence.

The danger is to arrive at an understanding which is wrong, which is easy to do with mathematics especially. However, if the theories support observation, that strengthens the case; and if there are no inconsistencies or paradoxes, that's even better. It will be a tough road because black hole science is already so full of inconsistencies, debates, partisanships, and so on, that forming a consensus that is aligned with reality will be no small feat. But, with persistence, hard work, and insight – anything is possible. Which leads to a final comment, which is: keep an open mind.

Chapter 3: Wormholes

The Earth-Moon system is very much a unit, but consideration of geophysics alone doesn't quite explain the relationship between the two bodies. Another avenue for understanding ranges into theoretical physics and wormholes. The Earth and Moon can be understood as two closed manifolds connected by a wormhole.

The Earth "environment" is defined as the surface of the Earth and the immediate space around it. This space can be thought of as geometrically closed, with the exception of the wormhole entrance, which leads to the closed geometry of the lunar environment. In general, the space around the Earth is like the inside of a sphere or more accurately, a toroid of revolution with a magnetic field. One can travel by spaceship around the Earth within this toroid. This is what the Space Shuttle did and what satellites do.

Entering the wormhole from Earth orbit may be as simple as launching a rocket into space. Once through, the negative energy of the wormhole starts to dominate the gravitational energy of the Earth. This changing dynamic takes the form of the stars being "extinguished" and a total blackness of space. The Apollo astronauts witnessed this during their journeys to the Moon. Since a wormhole is threaded with negative matter, this blackness can also be understood as *dark matter* which extinguishes the stars. When the astronauts reached the Moon, the stars came out again. As Apollo 11 rounded the Moon, Neil Armstrong reported: "Houston, it's been a real change for us. Now we are able to see the stars again and recognize constellations for the first time on the trip. The sky is filled with stars, just like nights out on Earth."

The exit of the Earth-Moon wormhole is, of course, the entrance to the lunar environment. When one approaches the physical reality of the Earth's interior, one is entering another closed geometry similar to the Earth's geometry, but in this case there is only the shining Sun, and instead of the Moon's image, its actual presence. As well, the image of the Earth as we know it shows up in the lunar skies, just as the Moon appears in the skies of Earth.

As one approaches the heavy gravity of a black hole, space has a tendency to decohere due to shockfronts associated with the huge gravitational forces associated with a black hole singularity. A wormhole causes spacetime to cohere, since the wormhole is the shortest path between its ends. When emerging from the wormhole and traveling into the Moon or Earth environments, spacetime decoheres again.

Moreover, the appearance of the Sun and the Moon in our skies can be understood as a decoherence of the wormhole itself. Likewise, in the lunar environment the blue orb of the Earth and the shining Sun appear there as a decoherence. Decoherence gives these images motion. One might say that the geometry of the skies is in motion from a stationary point inside of either the Earth or Moon environments. That is, the observer is stationary and the world moves around him. By contrast, the observer himself moves through a wormhole, and the wormhole geometry is stationary.

Inside of the Earth environment or the Moon environment the Sun, Moon, and/or Earth go through their normal courses. What about the stars? In the simplest scenario the stars are only visible outside of the wormhole. As one exits the wormhole, outside becomes inside, and so the stars are visible from the Earth environment and the Moon environment. But inside of the wormhole, the structure of the wormhole acts as a wall that's closed to exterior space. (A computer-based term for this is "firewall".)

The only way to travel outside of the Earth or Moon environment is to travel into the wormhole. The wormhole mouth is not a small opening somewhere but the whole expanse of orbital space between the Earth and Moon. One enters the wormhole via an orbital path called a *geodesic*. The wormhole is cohered and though space is dark, the traveler appears to leave the Earth behind, so at this level the wormhole provides an exterior view of the Earth, even though it's a view of the sphere of the Earth. But eventually even that disappears because the wormhole lens loses its focus. The motion of the Earth is superseded by the motion of the space traveler.

There is something more to all of this. The wormhole not only connects the environments of the Earth and the Moon, but may form the very fabric of the structure of the planets themselves. Dark matter may thread both planets, because the wormhole is their binding. Think of the Moon as the interior surface of the Earth. The electric charges Q and –Q which would seem to be in opposition to each other for two convex bodies, are actually in alignment if one of the bodies is allowed to be concave. The four-dimensional wormhole connection can be visualized so that the wormhole exit, the Moon, is a concave surface that forms the interior surface of the Earth, if one thinks of the Earth as a shell. From the Moon, electric charge –Q points into the mantle of the Earth and extends through to the surface of the Earth, which is convex. The field then exits the surface of the Earth as charge Q. In other words, the wormhole is in fact the physical structure of the Earth. Same for the Moon.

This amazing theory doesn't require the Earth and Moon to be one continuous body: they can be separate bodies. The main thing to consider is that the wormhole isn't just a four-dimensional bridge. It's the three-dimensional matter and the very fabric of the planets themselves.

Topology

The topology that connects the Earth and Moon can be characterized as a *genus 2 handlebody*, shown below. The simplest definition of genus g applies to an orientable unbounded surface, e.g. any closed surface in ordinary space that doesn't cross itself. Such a surface can be regarded as "a sphere with g handles." Thus, $g = 0$ for a sphere or any simply connected surface; $g = 1$ for a ring or torus; $g = 2$ for a double torus; and so on.

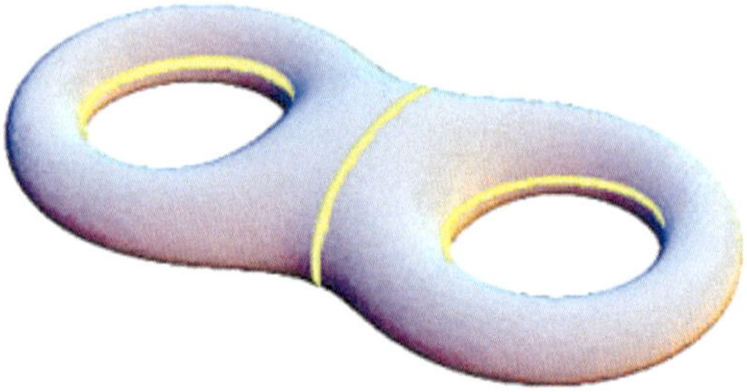

The wormhole of the simple genus 2 model must connect positive space with negative space. This positive-to-negative (or vice versa) requirement constrains the number of *coloring possibilities*. One can think of *color* as the labeling of a toroidal area as positive or negative. Therefore, the lefthand torus can be made positive, and the righthand torus can be made negative, or vice versa. Why can't they both be positive? The constraint is the lunar connector, which spans the middle portion of the handlebody, which defines the wormhole. It can't connect positive to positive or negative to negative. Likewise, the Earth on the left must be entirely positive; and the Moon entirely negative.

If a wormhole is allowed to connect positive to positive, or negative to negative, then, according to the strict rules of a wormhole, which says that one mouth has a positive charge and the other a negative charge, you wouldn't have a wormhole. So one must disqualify all configurations with a like-to-like connection.

The simple model doesn't handle the concentrism of the Earth and Moon. An embedded model does. This is more complicated because the embedding is reversible and symmetric. The wormhole connects an inner positive space with an outer negative space, or an inner negative space with an outer positive space. The main assumption is that a traveler in the lunar environment is outside of the Earth, which is an interior shell; and a traveler in the Earth environment is outside of the Moon, which is an interior shell. There are two great circles in the geometry: if one great circle is the equator, the other is a line of longitude around the torus, but when the geometry is reversed, the equator becomes the line of longitude and the longitude becomes the equator. So the whole thing reverses via a four-dimensional rotation.

Turning things inside out isn't a particularly realistic nor useful kind of mathematics. Ultimately it's a mind exercise; but there's another kind of math that is more useful and probably comes closer to the Earth-Moon reality. The idea is to classify closed manifolds according to whether they form together the boundary of an open manifold of the same kind. Two n-dimensional manifolds M1 and M2 are cobordant if there exists an open $(n+1)$-dimensional differentiable manifold whose boundary consists of the *disjoint union* of M1 and M2.

Imagine a model with three horizons: Earth, transition, and Moon, labeled as M, M', and M'' respectively. Let Bd mean "the cobordism of." Then the Earth and its transition region are defined as Bd N = M + M' , and the transition region and the Moon are defined as Bd N' = M' + M''. In the Earth case M and M' are cobordant and the unifying region is called N, which is "well-defined up to cobordism" (in the lingo of intersection theory). Likewise for N'.

Manifolds N and N' are four dimensional by the cobordism rules – which allows the traveler to jump over solid matter and move from interior to exterior. Symmetry around M' is expressed as Bd(N U N') = M + M'', where the union U is the manifold which results by gluing N and N' together along M'. Notice that M' has vanished.

Each statement like Bd N = M + M' is an *equivalence relation*. Since M and M' are disjoint, one can combine two statements as follows. Say that M is cobordant to M'. In fact we have the Earth statement: Bd N = M + M'. Say also that M' is cobordant to M'', or in Moon language: Bd N' = M' + M''. Then as we've seen before, but now putting it in slightly different language, Bd (N + N') = M + M' + M' + M'' or Bd(N + N') = M + 2M' + M''.

If M' is a closed manifold, and M' is an element of the set of all closed differentiable manifolds of dimension n, then M'~ is the *cobordism class* of M', and M'~ belongs to a cobordism *class group w* of dimension n.

The group *w* is a convergence class based on the *Thom group*, named after the inventor of cobordism.

The Thom group can be thought of as a sequence of projections from lower dimensions to higher, or vice versa. Eventually one reaches group equivalency, or isometry, if the dimension is high enough. As the cobordism converges the vectors associated with the manifold become more stable. These *vector bundles* are important because they're used to compute the polynomials that ultimately become gravity potentials. The *Thom group* stability dictates the stability of the associated vector bundles.

How do we know when the we've found the solution? The identity cobordism is Bd(N x I) = N + N = 2N. This means that the projection to an (*n*+*p*) space is the Identity in homogeneous coordinates. That's how you know you're done. In other words, the limit has been reached and intermediate terms have vanished. The the set of all closed differentiable manifolds of dimension *n* is an abstraction of the convergence group *w* of dimension *n* ; when 2N is cast as 2N~, i.e. put into the Thom group, we have 2N~ = 0 because nothing changes. Another way of saying this is that the 0-element of *w* is represented exactly by a closed cobordism, or by an isomorphism at the limit.

Thus *w* can be expressed as the limit:

w(n) = lim[p=infinity] P(n+p) (T(X(Gp)))

where *n* is the starting dimension of the cobordant manifolds; *p* is the addition to the starting dimension, representing *p*-planes in (*n*+*p*)-space; G is the Grassmannian algebra of *p*-planes; P is a projection; and the Thom space of the vector bundle X is *T(X(Gp))*.

Eventually one reaches group equivalency, or isometry, if the dimension is high enough. As the cobordism converges the vector bundles associated with the manifold become more stable. The nice thing about cobordism is that vector bundles start out relatively stable, so that manifolds can be approximated (if not solved) without going to the limit.

•

Let’s take a few steps back and reconsider the journey between the Earth and the Moon. One can hypothesize that the journey follows the path of an inversion process. If the Moon starts off as a torus that surrounds the Earth, with the Earth embedded inside of it, then after the inversion is complete the Moon will end up as a torus inside of the Earth.

Thus, the relative vantage point shifts from outside to inside, and vice versa. This means the Earth and the Moon, for lack of a better term, are *adnate*. They are conjoined. It doesn’t matter that they may be separated by 93 million miles or even 238,000 miles. The wormhole brings them together in holy matrimony (no pun intended). This may begin to explain why the Earth and the Moon can experience each other’s spacetime without suffering a time paradox. Time is wrapped up in the Earth and Moon as a unit.

Two tori that are adnate are both part of a *three-sphere*, called S3. Let D*n* be an *n*-dimensional disk, and S1 a one-dimensional sphere, or circle. Consider the decomposition of the hypersphere S3 = S1 x D2 + D2 x S1 . We find S3 by considering two full tori S1 x D2 = T’ and S1 x D2 = T”.

Consider a diffeomorphism Bd T’ to Bd T” which identifies the meridians of T’ with the parallels of T” and conversely. This is exactly the situation with adnate invertible tori, one describing the Earth and the other the Moon. One can write Bd (S1 x D2) = S1 x S1 = Bd (D2 x S1). The first S1 in S1 x S1 will be meridian in S1 x D2 and parallel in D2 x S1. For the second S1 in S1 x S1 the situation is reversed: it will be parallel in S1 x D2 and meridian in D2 x S1. In other words, the Earth meridian *m’* has the same orientation as the lunar parallel *p”*; and the Earth parallel *p’* has the same orientation as the lunar meridian *m”*.

As the astronaut travels moonward through the wormhole, space rolls back on itself with the astronaut on the leading edge of it. The critical piece of information here is that a star evolves as an observer moves through it. The astronaut continues to travel through the wormhole until the inversion is complete, with the new configuration forming the entrance to the lunar side, through which the astronaut emerges into the inside-out universe of the Moon. The return trip is just the same, except in reverse.

In this scenario, then, a wormhole entrance orbits the Earth's equator. Any geodesic trajectory to the Moon would be outward bound at the Earth's equator. The Earth environment, in the grand scheme of things, is one-half of the genus 2 universe. Kip Thorne discovered that if one end of the wormhole moves, such as in an orbit, while the other end remains fixed, this doesn't affect the distance between the two ends, if gauged by a tape measure that passes through the tunnel.

In ordinary space and time, the Moon orbits the Earth, but in a symmetric topology either can orbit the other. Even if the Earth and Moon are bound up in the dark matter of a wormhole, it doesn't matter which side is in motion. If the Moon side is fixed, and the other side orbits around Earth, the distance between the Earth and the Moon is always the same, and vice versa. Intuitively it might seem that the wormhole would become hopelessly tangled or coiled up, but the four-dimensional geometry of the wormhole prevents this, and that's pretty neat.

An *orbit* is defined by a group of rotations over the set of non-negative real numbers. The rotations are mod'ed by integer multiples of 2 pi. The integer-multiples are in the form of an integer *ring* denoted by the letter Z. An orbit (or fibration, or fibration bundle) can be defined by a function f(n) over the ring Z. Note that an orbit is not a closed circle and thus the radial notation *r* will be replaced with the designation *m*, meaning an orbit on the manifold M of Euclidean space E. Under the action of the group G = {f(n) : n in Z}, *m* is generated by f. The closure of *m* is a differentiable submanifold of E homeomorphic to a torus. The figure below represents something closely related to the orbit of f.

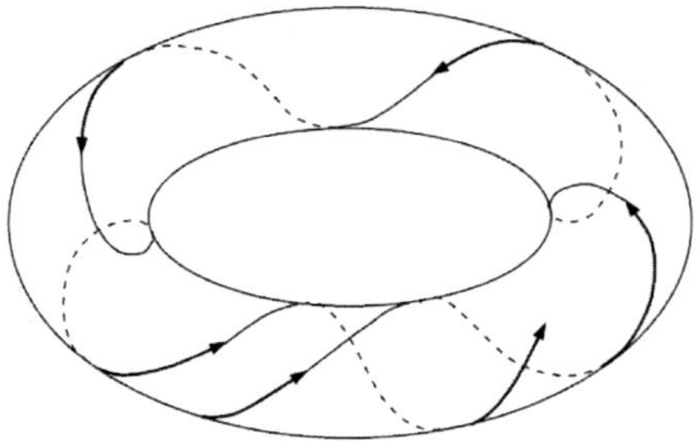

The figure represents only the topological structure of the orbit and its closure, not the orbit itself in four-dimensional Euclidean space.

Decoherence and Energy States

Decoherence is a mysterious process which happens in a symmetric way, depending on the location of the observer. When in the Earth environment, one can think of the wormhole as a decoherence; when in the wormhole, one can think of the Earth and Moon environments as decohered – or how else could they be turned inside out?

Decoherence is due to the enormous energies of a black hole singular region. From within the Earth environment, decoherence of the wormhole is somehow visible to us as the periodic orbits of the Sun and the Moon. And from within the wormhole, the decoherence of the Earth and Moon environments shows up as darkness.

Another way to think about the wormhole is as a set of energy states. The genus 2 universe, e.g. the interior environment of either the Earth or the Moon, is a ground state. Solar radiation that's expelled into space has a greater entropy relative to the more organized state of the Earth, which has a local entropy that's smaller. The Earth exists, locally, in the past relative to the Sun, and really has no memory of the Sun, since the Sun is radiating from a future place. This directly contradicts the belief that the Sun radiates from a primordial "center." (More on the "center" in the next section.) The Sun, compared to the Earth, has a greater entropy and the Sun as we see it thus occurs at a later quantum time, assuming that the Sun and the

Earth are part of the same quantum system.

Normally, an electron vacates an inner shell and excites an outer shell with its orbit in a higher state. Not so in the stellar framework. In the stellar framework a higher energy state can coexist with its past. In the future, there's a shell which has been vacated to a higher state, but when traveling back into the star, the shell becomes occupied again. This is because as one travels into a star the star evolves, or devolves as the case may be. The present and future show up together as, for example, the Earth environment (present), the image of the Moon (present), and the shining Sun (future).

An observer at the inner event horizon, or Cauchy horizon, of a black hole can look out onto flat de Sitter space and see the future of the outside universe. By extension, an observer would see the future of the black hole as well, including its radiation, perhaps focused by the lens of the wormhole interior. Therefore an observer on the Earth or Moon would see the Sun. An extension of a genus 2 model isn't necessary to model the higher energy state of the Sun. The Sun is a future state of the very same genus 2 universe that comprises the Earth and the Moon.

The Fourth Dimension

At first thought it might seem that it doesn't make sense for a wormhole to orbit any part of the genus 2 geometry. This intuition is correct in the sense that the entirety of the Earth and Moon is the entirety of the wormhole: there's no "doorway". The gravity of the middle portion of the wormhole, which is very strong and foreshortens space, would exceed the gravity of either the outer torus or the inner torus. Thus it seems unnatural for the wormhole to orbit the tori centers. Rather, the tori should somehow orbit or otherwise emanate from the gravity source at the center of the wormhole. But in a very real sense the tori *are* the gravity state at the center of the wormhole, which turned inside out, imparts motion.

In a nutshell, the concept of a geometric "center" doesn't apply to four-dimensional space. It only applies to three dimensions. We can see space in three dimensions, but we have no feeling for an added dimension. Where is the "center" in four dimensions? Faced with such a question, one is forced to make a relativity shift in the model assumptions.

The genus 2 universe is useful for modeling the spacetime of the Earth environment or the Moon environment as we see them. But the observer must be conscious of the fact that the images lensed by the wormhole are being lensed through a more complicated structure than the one most easily imagined by the apparent view due to energy states.

Obviously, the entire premise of a *connector* being a wormhole is flawed, since the wormhole geometry of a cylinder or tube is only valid for a two-dimensional sheet space. In other words, a three-dimensional cylinder can only connect two-dimensional sheets. Three-dimensional space must be connected by a four-dimensional wormhole.

One must cast off the limitation of a connector or tube with a two-dimensional cross section, and hypothesize that the wormhole mouths are the positive and negative (inner and outer) toroids of revolution *in toto*. The model is thereby demoted to a peripheral existence as a three dimensional cross section of a four dimensional reality harder to get at. The toroidal geometry isn't completely divorced from the four-dimensional world, since one can think of each torus as a hypersphere. If the wormhole mouths are the toroids themselves, then the intermediate space between the toroids can be thought of as the wormhole proper. But that intermediate space is the mantle of the Earth joined to the mantle of the Moon, which means that the wormhole is *everything*.

The wormhole-as-tunnel is merely a "dimension minus one" placeholder for a more complicated reality that can't be fully understood because we can't "see" four dimensions. But essentially both are equivalent. They do the same thing: the connector connects two-dimensional space, just as a four-dimensional superstructure which can't be directly experienced connects three-dimensional space.

Somewhere in that journey through space, from the Earth to the Moon, things get switched around in three dimensions due to an implicit rotation in four dimensions, but the origin of the rotation is four-dimensional. Mathematically, a four-dimensional rotation is sufficient to turn a three-dimensional projection inside out.

Ultimately, the concern over a geometric center is meaningless. After all, even in the conventional sense the space between toroids will follow the same lines of force as the toroids themselves. In this regard, the new model more easily accommodates a property that's missing from the conventional orbital model: *torque*. The concept of torque goes to the heart of an explanation of why the Earth and the Moon rotate in empty three-dimensional space, and more importantly, why the Moon's rotation is synchronous with its orbit around the Earth. As Einstein postulated, there are gravity potentials working behind the scenes. Working invisibly, they govern the dynamics and forces of the Earth-Moon system.

This idea of torque also has something to do with the geophysics of mantle convection, to be covered in the next chapter.

Chapter 4: The Core

Spherical Harmonics

In 1966 Miyamoto wrote his paper on the spherical harmonics of Mars, when the only space-based data that existed for Mars was from the Mariner IV orbiter, which reached Mars in 1965. (Mariner III had failed during launch, and Mariner II was sent to Venus.) Thus the physical data was fuzzy at best, but much better than visual observations from Earth. This fuzziness gave Miyamoto a measure of freedom to interpret results that may not have been possible with later, more accurate and distracting topographic data. This scientific freedom resulted in a superb interpretation of the Martian surface.

Miyamoto assumed that magmatic differentiation had taken place on Mars, and that the crust was composed of two layers, the basic maria and the more acidic continental deserts. He assumed the same thing for the lunar surface, and of course by 1966 the theory of continental drift on Earth had become established. Thus, mantle convection on all three planets was assumed to be responsible for the distribution of maria and continental or highland areas. Continental blocks tend to shift towards descending currents of convection, and the maria spread over the ascending currents.

The most basic spherical harmonic is of order 1. One can picture a descending current on the Moon's far side that passes through the center of the planet and returns to the upper layer on the near side, and then passes below the crust along the full circumference of the sphere before subsiding again at the far side or antipode, towards the center. This type of mantle convection is characterized by a continental hemisphere and an oceanic hemisphere. The ur-continent of the Earth, before the drift of its continental fragments, corresponds to this initial, order 1, stage. In general, the topography of Mars also has this quality, with its north polar ocean and its south polar highlands. The Moon's near side can be interpreted as a pole facing the Earth, of ascending currents in general; and its far side as the opposite pole characterized by highlands and descending currents.

This also jibes with the meridian/parallel switch of the Earth/Moon topology outlined in Chapter 3.

In a classic bit of scientific reasoning, Miyamoto hypothesized that Mars conforms to third-order spherical harmonics. The annotated MOLA map below shows longitudinal variations corresponding to third-order spherical harmonics. In general there's a tri-axial arrangement emanating from either pole, indicated with yellow arrows in the map.

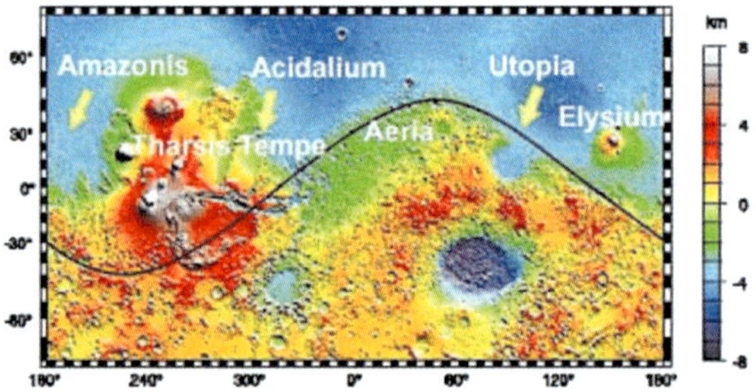

There is also a latitudinal regime for such harmonics. Some researchers prefer to align the northern and southern hemispheres of Mars with the geomagnetic equator rather than the geographic one. Thus the black sinusoidal line shown in the map above divides the Martian surface into roughly two hemispheres. To the south of this line the altitudes are on average higher than to the north. Because the dividing line is inclined with respect to the equator, the hemispheres can be referred to as "northernly" and "southernly" in contrast to "northern" and "southern". However, Miyamoto doesn't resort to this device and considers the Mars hemispheres directly.

All three planets went on to develop higher-order harmonics. The Earth evolved into a fourth-order system, and then was characterized by fifth order harmonics starting about 200 million years ago. The Moon seems to have frozen in "state" early in its evolution, but did develop along half lines with second order harmonics, with descending currents at the libration points on either side of the Moon's face, and an ascending axis towards the Earth.

Mars evolved into a system with third-order spherical harmonics. Miyamoto came to this conclusion from albedo observations. The snow line around the south pole appears more irregular than that of the northern cap, and isolated mountains (Novus Mons and Mons Argenteus) appear to be covered with snow after the snow line has retreated polewards. Thus the south pole appeared to Miyamoto to be highlands – and he was right.

On the other hand, the north pole was observed as three conspicuous maria – Mare Acidalium, Mare Utopia, and Propontis (Mare Amazonis) – all between northern latitudes of 30 to 45 degrees. (See the map above.) North of these maria, the snow line of the polar cap is more circular than that of the southern cap, and Miyamoto concluded, rightly, that the north polar region is flatter and distributed with maria.

This then describes, so far, the classic distribution of an order-1 spherical harmonic. An order-3 surface has an additional belt of maria in the southern hemisphere, and continental features in the north. Thus, Miyamoto singled out Hellas, Noachis, Argyre, Phaetontis, Electris, Eridania, and Ausonia (some of these are classically named albedo features, and some are now known to be craters or maria) in the southern latitudes between 30 and 55 degrees. In the north, continental features singled out by Miyamoto, between northern latitudes 0 and 30 degrees, are the Tharsis-Tempe region (to the west of Acidalium); Aeria (an old name for "Egypt" and now consolidated with Arabia Terra, to the west of Utopia); and Elysium (to the west of Propontis/Amazonis). Thus, an order-3 surface is characterized by four latitudinal bands of alternating continent and maria rather than two for an order-1 surface.

As well, an order-3 surface has longitudinal variation, as characterized by the tri-axial arrangement of the Martian north pole (as in the figure above). There are three so-called "vapor courses" connecting the two hemispheres. Miyamoto gives their locations in a rough way as: Mare Acidalium-Nilokeras-Lunae Lacus-Sinus Aurorae (shown with a yellow arrow under "Acidalium"); Utopia-Nepenthes-Syrtis Major (shown with a yellow arrow under "Utopia"); and

Propontis-Cerberus-Mare Cimmerium (all of these names have passed out of use except for Terra Cimmeria on the southernmost point: the area corresponds to Arcadia and Amazonis Planitia in the north, shown with a yellow arrow under "Amazonis").

Therefore the tri-axial arrangement of the north polar area is in fact an integral part of an order-3 spherical harmonic surface. The phrase "vapor course" can be explained by the old-fashioned idea that the darkening of canals or tectonic valleys from season to season was due to the melting of ice in the northern hemisphere, and the movement of water vapor or clouds from the northern hemisphere to low latitudes of the southern hemisphere. This idea in fact still has merit, given the scenario of ice clouds being one source of solar reflections.

Interestingly, the order of harmonics of a planet's surface directly correlates to the size of its core. An order-1 surface has a minimal core or no core at all, allowing the mantle currents to pass virtually through the center of the planet. The core size favorable for order 3 convection on Mars is between 0.35 and 0.50 of its radius. Thus Mars is thought to have a small iron core. The Earth's order-5 harmonics require a core larger still. The Earth's mantle stops at about .62 of the Earth's radius. Von Weiszacker proposed a concentric system of vortices, not dissimilar to currents of mantle convection, for the structure of the inner solar system. The concentric system might also describe piggy-backed concentric convection cells, with higher orders residing farther out from the order-1 core.

Though there is plenty of evidence for the Moon being the inner surface of the Earth, a second inner region for Mars has no support whatsoever. One would have to combine such a theory with many other assumptions which are hard if not impossible to prove. Mars is in its own system.

Miyamoto calculated the temperature difference between ascending and descending mantle currents on Mars as 0.5 deg C. Ascending currents, trapped inside the hot rock of the mantle, and uncooled by continuous exposure to the crustal regions, is hotter than descending currents. For the Moon, and assuming an order-2 convection,

Miyamoto came up with a temperature difference of 50 deg C. This isn't because the lunar surface is Sun-baked, but because things generally are hotter in the lunar mantle compared to Earth. The Earth weighs in with a temperature difference of 0.3 deg C. The table below shows this sequence along with the mantle current-velocities and core sizes for the three planets.

Planet	Temp diff	Mantle velocity	Core size
Moon	50 deg C	2 cm / year	< 0.3 radius
Mars	0.5 deg C	25 cm / year	0.3 - 0.5 radius
Earth	0.3 deg C	100 cm / year	0.62 radius

The takeaway from this data is that, if the Moon really is the inner surface of the Earth, the convection of the lunar mantle is both hotter and also moves much more slowly than the convection on the Earth side. This makes sense as it's known that convection on the Moon came to a halt at the same time the developing harmonic scale stopped at 2. But apparently there's still very slow convection, stirred up perhaps by the fast-moving currents on the Earth side. And yet, where is the proof?

The Japanese Study

With the new understanding derived from diamonds as described in Chapter 2, and new technologies such as high-tech pressure anvils, more observational data is becoming available about the interface between the Earth's mantle and the core. In 2013 a Japanese team published a study which supports the idea that the speed of convection at the mantle/core boundary is indeed much slower than above.

Part of the water existing in abundance near the Earth's surface is transported to the Earth's mantle, extending from a depth of 30 to 2900 kilometers, as hydrous minerals by subducting slabs. Hydrous minerals named "phase D" have been observed in the lower mantle, marked between the depths of 660 to 2900 km. However, it has been reported that phase D dehydrates and decomposes at pressures near 44 GPa, which corresponds to a depth of 1250 km, and that no water can reach the deeper regions of the lower mantle.

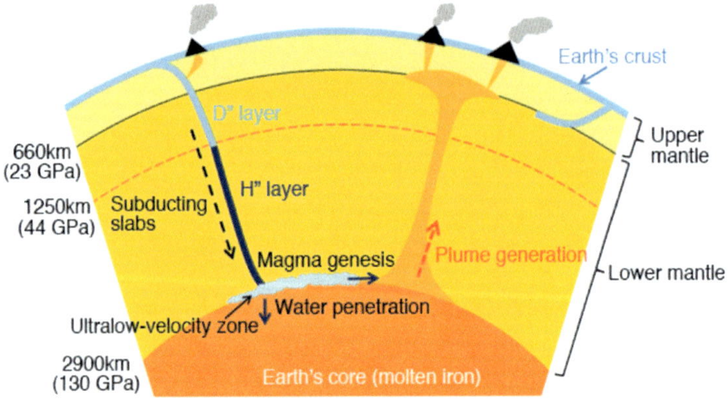

Jun Tsuchiya predicted that phase D transforms to different crystal structures at pressures near 44 GPa based on theoretical calculations. To verify the prediction of Tsuchiya, the research group carried out experiments using an ultrahigh-pressure apparatus and *in situ* synchrotron radiation experiments. As a result, they found that phase D transforms to new high-pressure hydrous minerals named phase H near the theoretically predicted pressure. The crystal structure of phase H was slightly different from that predicted by Tsuchiya but very similar.

Phase H is likely to be stable even in the mantle-core boundary. It seems to have significant effects on the composition and dynamics of the deep Earth, for example, the general circulation of water in the deep Earth, the generation of upwellings (plumes) in the mantle-core boundary, and the penetration of water into the molten iron, which is the main component of the Earth's core.

Earth consists of three layers, namely, the thin crust extending to a depth of approximately 30 km, the mantle extending from a depth of 30 to 2900 km, and the core from a depth of 2900 to 6400 km. The mantle mainly consists of rocks such as peridotites, whereas the core mainly consists of iron.

The mantle is divided into three zones, namely, the upper mantle extending from a depth of 30 to 410 km, the mantle transition zone from a depth of 410 to 660 km, and the lower mantle from a depth of 660 to 2900 km. The largest zone is the lower mantle, which makes up 60% of the volume of the Earth. The depths of the lower mantle are in contact with the Earth's core.

Plumes are high-temperature upwellings rising from the deep mantle, in contrast to subducting cold slabs and mantle materials. Seismological studies have shown that huge superplumes originated from the core existing under Africa and the lower part of the Pacific Ocean. Rocks may partially melt in the region where plumes are generated. The presence of water is an important factor in the generation of plumes because the temperature at which rocks melt is decreased with the presence of water.

The critical information for the Earth-Moon structure is, what is the speed of the mantle-core interface? The region in the lowest part of the mantle near the mantle-core boundary, where seismic waves travel at a very low velocity, is called the *ultralow-velocity zone*. The ultralow velocity seems to result from the reaction of the mantle composed of rocks with the molten iron or the partial melt of mantle materials. The presence of water is an important factor in the formation of such a low-velocity zone (and the generation of plumes) because, as already mentioned, the temperature at which rocks melt is decreased with the presence of water.

So water not only exists in a vast reservoir beneath the Earth's surface, but is the key to the dynamics between the Earth and its core – and by theoretical extension, the Moon. Hydrous minerals, such as serpentine, contain hydrogen as a major component. The magnesium-rich hydrous minerals formed at high temperatures and pressures in the Earth's interior are called high-pressure dense hydrous magnesium silicates (DHMSs) or *alphabet phases*, which are considered to be transported to the deep Earth by subducting slabs.

Phase D is the hydrous mineral considered the only DHMS in the lower mantle. Since the discovery of phase D by an Australian researcher in 1986, there have been reports of the discovery of phases

E, F and G. However, subsequent studies have revealed that phase E exists only in the upper region of the mantle and that phases F and G are the same as phase D.

Phase D seems to transform to the new hydrous mineral phase H within the slabs subducting into the lower mantle and transports water near to the Earth’s core. The pressure and temperature ranges of stable phase H are significantly widened by the addition of aluminum (Al), which is an important component of phase H next to magnesium (Mg) and silicon (Si).

Water on the Moon

If the Earth and Moon are connected by the mantle, then there must be some evidence of water in the Moon’s mantle. In 2013 researchers discovered that the interior of the Moon contained water early in its history. “The data suggests that the formation theories of the Moon may have to be rethought.” University of Michigan researchers and colleagues found tiny amounts of water in mineral grains from samples from the lunar highlands obtained during the Apollo missions.

The lunar highlands are thought to represent the original crust that crystallized from a mostly molten early Moon, suggesting the lunar interior contained water during this molten stage before the crust solidified. That’s difficult to reconcile with the theory that the Moon formed from the debris generated during a giant impact with the proto-Earth and any water in lunar rocks was added after its formation by smaller meteorite impacts or the solar wind.

“Because these are some of the oldest rocks from the Moon, the water is inferred to have been in the Moon when it formed,” UM researcher Youxue Zhang said. “That is somewhat difficult to explain with the current popular Moon-formation model, in which the Moon formed by collecting the hot ejecta as the result of a super-giant impact of a Martian-size body with the proto-Earth. Under that model, the hot ejecta should have been degassed almost completely, eliminating all water.”

However the “Genesis Rock”, a lunar sample collected during the Apollo 15 mission and thought to be a piece of the Moon’s primordial crust, contains traces of water. The new findings indicate that the early Moon was wet and that water there was not substantially lost during the Moon’s formation. It means that the Moon’s mantle contained water that made its way into differentiated rocks formed by mantle convection starting 3.8 billion years ago.

Thoughts

It’s interesting to play the math game. The core boundary starts at 2900 kilometers. This translates to 1802 miles. Subtracting this from the Earth’s radius of 3440 miles leaves 1638 miles. Now, if we assume that despite the distortion of spacetime the inner core is the size of the Moon, with a radius of 1088 miles, then the depth of the lunar mantle, up to the Earth-core boundary, is 550 miles or 885 kilometers. This agrees almost perfectly with the fact that large-scale melting of the Moon’s mare basalts, at depths of 200 to 500 kilometers, got started some 3.8 billion years ago. It leaves room for a deeper, more mysterious region that forms the interface with the Earth’s mantle, and is the very substance of the Earth’s hot core.

Chapter 5: Tectonics

UFOs and Tectonics

Mars and Earth orbit around the Sun in a repeating cycle. Every 779 days the Earth, Sun, and Mars line up so that the Earth is exactly on the line between the Sun and Mars. This is called the Earth-Mars opposition. During a two-month window during the opposition the magnetic fields of the two planets are aligned in such a way that UFO sightings increase dramatically.

This relationship between Earth and Mars is reminiscent of a tuning knob on a radio, in that a frequency tuned in by the knob will become clear if the knob is in the right place, and then fade out to static when the knob is in between radio stations.

The theory is that the Earth, like a radio receiver, picks up signals from the nearby stars, due to an interaction between any of those stars and our own Sun that mimics the interaction between the Earth and Mars on a grander scale. As stars come into alignment the radio is tuned in.The idea is that stellar alignments bring images from stellar distances to the skies of Earth.

UFOs look like solid objects. They look three-dimensional. For three-dimensional signals to be received in our atmosphere, the radio waves would have to travel through four dimensions. The analog is radio waves that travel through three dimensional space. Upon being received, the decoded radio signal is two dimensional. If the Earth is a black hole, then four dimensional radio waves are always traveling through the Earth and through our bodies, but only manifest in three dimensions when the conditions are right, which, given the frequency of UFO sightings, is rarely.

The radio wave is brought down from the four-dimensional aether as a three-dimensional object, though it's still only a signal. The military and CIA have a name for these signals: "phantoms" because they show up on radar but have no mass.

Major General John A. Samford, Director of Air Force Intelligence, at a Pentagon press conference concerning UFO sightings over Washington DC, held July 29, 1952, ruled out guided missiles as being mistaken for UFOs. In a long discussion Samford reduced the saucers to something with unlimited power and no mass. “You know what no mass means,” he said. “There’s nothing there.”

The theory is that a certain class of UFO is a three-dimensional signal that has been captured by the alignment between the Earth, the Sun, and a third stellar object. The UFO doesn’t really exist on Earth, but like television in two dimensions, has been transmitted from afar. A regular radio signal may originate in Chicago and be received one hundred miles away in Rockford. The radio personality sits in a studio in Chicago, and the listener in Rockford hears her voice. Likewise, machines flying in the atmosphere of a nearby star do not leave that star, but we see the phantom image of that machine on Earth.

In 1949 Project Sign and the Air Force believed that 22 nearby stars had orbiting planets. The 1949 report stated: “Outside the solar system other stars – 22 in number – have satellite planets. Our sun has nine. One of these, the Earth, is ideal for the existence of intelligent life. On two others there is probability of life [Mars and Venus, they thought]. Therefore, astronomers believe reasonable the thesis that there could be at least one ideally habitable planet for each of the 22 other eligible stars.”

One of the eligible stars mentioned in the report is Wolf 359, about 8 lightyears from Earth. Other nearby stars include Sirius, 9.11 lightyears distant; Alpha Canis Minoris, 10.22 lightyears away; Procyon, 10.4 lightyears away; and Kruger 60, 12.62 lightyears from Earth. Two others (distances not specified), 61 Cygni and 70 Ophiuchi, were known to have planets. Today there are literally over 100 stars known to have orbiting planets, including Alpha Centauri, 4.29 lightyears from Earth, which in 2012 was found to have at least one orbiting planet.

Scientists with Project Sign postulated that a radio connection could conceivably link Earth to a nearby star in both directions, so that a civilization on a nearby star "might observe on Earth that we now have atomic bombs and are fast developing rockets. Since the acts of mankind most easily observed from a distance are A-bomb explosions, we should expect some relation to obtain between the time of the A-bomb explosions, the time at which the space ships are seen, and the time required for such ships to arrive from and return to home base."

This scenario is close to the theory being proposed, except that no physical hardware travels between the planets: only radio waves. The time paradox of black-hole theory, by which the signal traveling near the speed of light experiences a time foreshortening, would not apply to this theory of interstellar radio, since people on Earth remain stationary in relative terms, and so experience no time compression. We see the signal in the time it takes light from the star to reach the Earth, even though the signal itself may be younger than its travel distance.

Carl Sagan proposed the actual medium through which such signals would travel: hydrogen.

"There is a question of frequency. What channel would you listen on? At the low frequencies there is a background from charged particles in magnetic fields in the Galaxy, the galactic background. It's noise. And it gets to be very substantial noise. This is not where you would want to transmit or receive. At the high-frequency end, there is another source of noise, intrinsic to the quantum nature of radio detectors. In the middle there is a broad region where the noise is low, and this is the window in which it makes sense to transmit. In this window there are certain spectral lines, for example, of atomic hydrogen, the most abundant atom in the universe."

Certainly hydrogen only exists in three dimensions, so that transmission through the fourth dimension might use hydrogen or hydrogen products like methane or water as conduits. The Sun is mostly hydrogen. If a black hole in its entirety is a radio detector, then the quantum nature of the detector would be the quantum nature of the Sun itself – and this may start to explain how signals are received, that is, when the Earth's higher energy state, the Sun, somehow aligns with some other planet's higher energy state, being another Sun.

UFOs maintain a radio silence, though their appearance causes radio static and whistles and other noises. Nothing inside the craft communicates with the Earth. In 1949 the Air Force was able to identify classes of UFOs. For example, night sightings were called "Group 2" and included "Type 4," described as "a machine with rotating red-green-white lights and fixed white beams. May be a rotating disc type. Recorded speed, well over 1,000 m.p.h. Speeds estimated by competent pilots."

Night sightings included UFOs with colored, revolving lights; and with lights that blinked at intervals. The lights could have been an attempt to communicate, but the radio silence made that unlikely. Intelligent beings who had mastered space travel would certainly be able to duplicate our radio transmission system, but no strange codes or unfamiliar word-sounds had been heard.

Daytime sightings were classed under "Group 1": "The most numerous reports indicate daytime observation of metallic disclike objects, roughly in diameter ten times their thickness. Some suggest the cross-section is assymetrical and rather like a turtle shell. Reports agree that the objects are capable of high acceleration and velocity. They are often sighted in groups, sometimes in formation. Sometimes they flutter."

One reason contact with radio phantoms has never been made is because the images are ephemeral. The occupants are presumably going about their business oblivious to the fact that their spacecraft is being broadcast to a million other worlds. Likewise, a pilot for Pan American Airlines is oblivious to the possibility that a three-dimensional ghost-image of the 757 he's piloting may be showing up in the skies of Alpha Centauri planet A.

A radio wave traveling through a wormhole in four dimensions must have a very real connection to the wormhole's structure in three dimensions, being an entire world. What is the mechanism by which the signal is boosted into the fourth dimension and spun down again to the third on the Earth side of the radio channel? An alignment of stars forms an electromagnetic bridge. In one sense, the four-dimensional fabric of spacetime is the very fabric of the planet itself, so that one can think of Alpha Centauri and the Earth as inhabiting the same space, temporarily, linked by the stellar alignment. In three dimensions they're 4.29 lightyears apart, but if the timing is right, the atmospheres of the two stars can overlap in a tenuous radio communication. In this way, the three-dimensional image of an alien craft is "brought down from the aether." If the transmission is weak, the UFO would seem to appear and then fade, flutter, or vanish.

Faults

A solar opposition may put stresses and strains on the Earth because of the slight torque exerted on the Earth by the Sun on one side and Mars on the other. The rotational equilibrium would be disturbed slightly, thereby activating faults, which respond by producing electrical currents which, under rotational forces, are transmitted between, say, Earth and Mars, or Earth and the Moon.

In March 1950 Captain Jack Adams and First Officer G.W. Anderson, both of Chicago and Southern Air Lines, were *en route* from Memphis Tennessee to Little Rock, the first leg of a regular flight to Houston Texas, when they encountered a UFO over Stuttgart Arkansas, not far from Marianna.

In 2009, the Associated Press released a story which can be summed up as follows. Haydar Al-Shukri, the director of the Arkansas Earthquake Center at the University of Arkansas at Little Rock, discovered a previously unknown fault in eastern Arkansas that could trigger a magnitude 7 earthquake. He said the fault is separate from the New Madrid fault, the principal fault of the region. The New

Madrid fault was responsible for a series of three magnitude-8 earthquakes in the winter of 1811-12 that caused the Mississippi River to flow backwards.

Acres of cotton fields cover the fault west of Marianna, about 100 miles east of Little Rock, but stretches of fine sand mixed with fertile soil gave away the fault's location. Al-Shukri said that liquefied sand bubbled up through cracks in the earth. Ground radar and digs revealed vents that let the sand reach the surface.

The fault was likely created in the last 5,000 years and sparked at least one magnitude 7 earthquake in its history. Such temblors leave massive destruction in their wake. Al-Shukri warned, "This is a very, very dangerous area at risk of an earthquake." He added that such a quake would affect Little Rock and neighboring states such as Tennessee and Mississippi.

Hundreds of earthquakes occur each year, including several in Arkansas. Most Arkansas earthquakes occur in the state's northeast corner in the New Madrid Seismic Zone, but Al-Shukri said the Marianna fault was not connected because it was too far from the temblors caused by that zone.

Stuttgart, the scene of the UFO sighted at 2000+ feet, is about 50 miles from Marianna.

A comprehensive study completed in 1982 by Lee Trotta highlighted several fault lines traveling through several regions in Wisconsin. In the far northwest areas of the state, there are several fault lines including the Lake Owens Thrust Fault and the Bois Brule River Fault. The center of the state from southwest to northeast is clear of fault lines. From the far southwest to the east central and southeast, there are again several fault lines, including the Cottage Grove Fault between Madison and Milwaukee. There is one specific fault near Green Bay Wisconsin called Dutchman Creek Fault. This fault is a 60 feet offset type, close to the town of Ashwaubenon, just south of Green Bay.

Apparently these faults are not active, but the Earth and Moon have electromagnetic "hot spots" and, for whatever reason, the photograph below indicates that the Green Bay region was a hot spot in 2007.

A good candidate on the local level for, if not the medium, then at least the catalyst for UFO activity, are fault lines. Earth has many fault lines that have generated UFO phenomena observed during the decades of the 20th century. Even subduction zones in the ocean seem to generate "unidentified submersible objects." More, UFOs have been observed to fly into the ground and vanish without even raising a cloud of dust. California and its coastline has been the site of more UFO sightings than any other state in America, possibly due to the San Andreas fault and the crustal subduction zone off the coast.

The science of seismology progressed only after the famous 1906 San Francisco earthquake. A key figure in its advancement was the great Berkeley geologist Andrew Lawson. Before 1906, Lawson had done some field work but his understanding was limited. In 1893 Lawson took a steamer trip from San Diego to San Francisco, stopping off to examine coastal formations at several places. He noted that geological features in San Benito (southern California) and Santa Clara (northern California) counties, as well as the San Francisco Bay valleys, were remarkably arranged in a line.

In 1895, more than a decade before "the Big One," Lawson took the name of the San Andreas Lake, so christened in 1774 by a Spanish

Padre named Francisco Palou, as the name for the crack in the earth spanning the distance from northern to southern California: the San Andreas Fault. And yet Lawson couldn't bring himself to fully embrace the enormity of the distance – some six hundred miles – and so limited his discussion of the topic to the narrow valley on the San Francisco peninsula that holds San Andreas Lake and Crystal Springs Reservoir – not much more than a dozen miles in length.

When the San Francisco earthquake hit in 1906 its destructive power was a wake-up call. Until this time geologists weren't sure how earthquakes and faults were connected. Many believed that faults were the byproducts of earthquakes, not their source – although the two are so closely related that precedence can't really be established between one and the other.

After the quake, Lawson and a team of colleagues decided to reevaluate the California fault system. They walked the "mole tracks" (the fault lines) that had ruptured across barnyards and meadows in northern California. Then they continued south for six hundred miles, reading the landscape, discovering new unbroken sections of the faultline. The fault kept going and going to Los Angeles and beyond. In 1908 the team published the Lawson Report, which showed this rip in the Earth in vivid photographic detail.

Today, scientists know that faultlines are partly the result of stresses and strains in the planet's rotation about its polar axis. Most faults run north to south, reflecting the polar alignment of the planet's spin. The same fault orientation, along the polar axis, occurs on the Moon and Mars. Scientists also know that mountain chains on the western coasts of the Americas follow a course from north to south, and are known to be the sources of gravity anomalies. In the same way that faults and earthquakes are codependent, gravity anomalies can't claim priority over the very presence of mountains. Mountains and faults are so similar that gravity waves are surely associated with faults.

At first Lawson didn't fully appreciate the link between the San Andreas fault and earthquakes, but time and experience remedied that. Today we aren't fully understanding the link between faultlines –

such as the Crystal Springs Reservoir span of the San Andreas fault – and the production of UFOs, or, perhaps more precisely, the leaking of an alternate reality into the reality of Earth.

Inactive faults can be found almost anywhere on this planet if one digs deep enough. The radio theory can therefore be applied, with enough research, to almost any location on Earth. Because of its predominance everywhere, geology is a weak correlate. The same reasoning applies to UFOs themselves, which have been witnessed in almost every location on Earth at one time or another.

Still, one more example fills out the picture a bit more. In the early 1980s the Defense Mapping Agency carried out gravity surveys west of Rachel Nevada near Area 51. (Yes, *that* Area 51!) Says someone involved: "The nature of our work required that we measure gravity anomalies over several miles and loop them back. For the uninitiated, gravity is not a constant, it varies over terrain and oceans. We did this type of work in Nevada, and parts of California that we tied into a grid that was known at the time as The Land Region Gravity Model or LRGM. We were told that this data was to be used for Reagan's MX-Missile program."

The program ended in the early 1990s. It turns out that the area around Area 51 is a massive crater, probably an impact crater that created a crack in the surface of the Earth that released volcanic rock. Area 51 is at its exact center. The so-called Alamo bolide impact occurred 367 million years ago, when one or more hypervelocity objects from space slammed into shallow marine waters at a site that is now the Devonian Guilmette Formation of the Worthington Mountains and Schell Creek Range of southeastern Nevada.

However the sheer size of the ancient and buried crater, extending from Las Vegas in the south to Eureka in the north, suggests that the crater is volcanic. And its location at the southern end of the Sierras, and underneath a continent, suggest that it's the surface outcrop of an ancient plume. Although gravity at the Earth's surface is very nearly constant, it's slightly greater where dense rock formations lie close to the surface. Gravitational force, therefore, increases over the tops of anticlinal (arch-shaped) folds and decreases over the tops of salt domes.

Surveys can be conducted to measure the gravity of such geological features as earthquake faults and ancient meteorite impact structures. Relatively positive anomalies may indicate that rocks of higher density were uplifted during the cratering process to form the uplifts and rings. Rock fracturing and low-density impact melt-rocks, suevites and other breccias (mixed, crushed rock) cause negative anomalies.

Clearly Rachel Nevada is the epicenter of either an ancient impact crater or a massive upwelling. The government was mapping gravity anomalies around the crater in the early 1980s. UFO researchers have made films that prove that Area 51 is a hot spot of UFO activity. One can conclude that this activity isn't related to a secret cache of flying machines kept by the government, but to the natural geology of the area, e.g. stresses and strains combined with natural faults and electromagnetism associated with the crater.

Thoughts

The pairing of UFO sightings with geology, each comprising a vast dataset, doesn't necessarily prove anything. Thinking along a different line, geology is a long-term feature, while UFO sightings rarely last more than an hour. Again, a comparison of something fleeting with something that endures over epochs is not a strong case for a correlation. However multiple sightings at the same location would strengthen the correlation.

The last cautionary word goes to Carl Sagan:

"There is a stunning correlation between earthquakes in the Andes and oppositions of the planet Uranus [every 370 days]. Is this a causal connection or not? First thing you ask is, how many connections had to be looked for before this particular one was derived? Volcanoes in Sicily with oppositions of the planet Mars [every 779 days] – think of how many volcanoes there are in the world, how many earthquakes there are, how many planets there are, how many stars. If you start making a specific number of cross-correlations you will, of course, on occasion, come upon a coincidence. And what you have to do in *a posteriori* (after the fact) knowledge is add up all those other cases of possible coincidences that you looked at or could have looked at. You may end up with a slim reed upon which to base your conclusions."

Chapter 6: The Van Allen Belts

Things can move outwards or inwards, and so there's a fifty-fifty chance that a correlation can be drawn between a black-hole event horizon dynamic and an observation of some dynamic of the Sun, the Moon, or the Earth environment. There's a fifty-fifty chance that *anything* can be correlated to the black-hole dynamics. One must therefore ask: are the observations somewhat related? Are we throwing apples into the barrel with oranges, or apples with apples? If the latter, then the theory would tend to become stronger as more correlations are found without creating any contradictions.

Scientists already agree that there's a relationship between the Van Allen belts and the South Atlantic Anomaly (SAA). Apples and apples. But things are sketchy when it comes to the Sun's magnetic field and the Van Allen belts. And things are really sketchy when considering the Earth and Moon as a unit interior to the Sun! However, all of these things are related. The Earth feels the effects of solar flares, for example, as do the Van Allen belts and the Moon.

The South Atlantic Anomaly is an area of heavy radiation associated with the inner Van Allen belt. Scientists believe the Earth's magnetic field is somehow involved, and this further points to lunar analogues, since the Earth and the Moon form a unit. The drift rate of the SAA is very close to the rotation differential between the Earth's core and its surface, estimated to be between 0.3 and 0.5 degrees per year to the West. However this doesn't make sense if the core is the Moon, which is fixed relative to the Earth in four dimensions. Likewise, the libration points of the Moon are fixed because the Moon always faces the Earth, and though the libration points librate, the effect is small. The observed drift is probably the drift rate of the Van Allen belt itself, which is a synchrotron and subject to all kinds of magnetic forces.

Over time, the Van Allen belt has drifted, lost magnetic field strength, and dipped closer to the Earth, accounting for the teardrop shape of the SAA. However, the ongoing dipole/quadrupole interaction between the Earth and the Moon can't be ignored, and this is probably cyclic. The leading edge of the SAA dewdrop is aligned with the direction that the Sun is perceived to travel across the Earth, even though the Earth rotates to the East. Thus the Van Allen belt is slowly rotating in the same direction as the Sun's apparent motion: West.

However, a counter-rotation of the Van Allen belt relative to the quickly rotating Earth doesn't make much sense. Rather, the rotation of the Earth probably affects the fixed magnetic field of the Van Allen belt. The Van Allen belt is inclined 11 degrees to the Earth's poles, corresponding to the magnetic poles of Earth and to the geomagnetic equator. One "side" of the belts is some 1200 kilometers above the Earth's surface, while the other "side" is only 500 km high and creates the SAA.

The Van Allen belt seems to absorb the magnetic fields of the Sun when the Sun appears to lose its magnetic field during its pole switch every 11 years. The flip is a short-term manifestation that may show up as an extra Van Allen ring, positioned between the inner and outer rings and lasting for about a month. This is shown by the middle red ring below.

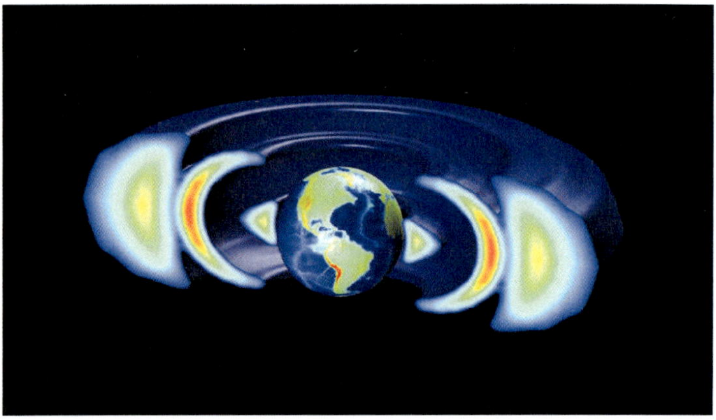

The Van Allen belts can be thought of as the Earth's convective envelope – a term usually associated with the Sun. Just as the Sun's core uniformly rotates within the Sun's convective envelope, so too for the Earth within the Van Allen belts. Since the SAA is limited to one Earth location it must be a projection from within or without – probably both. The magnetic lines of force are therefore distorted by the Earth's rotation.

Van Allen Probe

The rotation of the Earth was previously considered incapable of affecting the motion of radiation belt particles, which have velocities approaching the speed of light. Solar wind activity is the primary force behind changes in the radiation belts. Radiation levels across the belts are affected by solar activity that causes energy and particles to flow into near-Earth space. However, in 2014 scientists uncovered a mysterious structure they called *zebra stripes*. The stripes extend through the inner radiation belt, moving as the Earth rotates, and are visible even during low solar wind activity.

Their theoretical modeling clearly identified Earth's rotation as the mechanism creating these patterns. Because of the tilt in Earth's magnetic field axis, the planet's rotation generates an oscillating, weak electric field that permeates the entire inner radiation belt, extending from about 500 miles above the planet's surface up to roughly 8000 miles.

In 2012 NASA launched twin satellites called the Van Allen probes to study the Van Allen radiation belts in detail. Data from the Van Allen Probes Ion Composition Experiment (RBSPICE) on board each of the twin spacecraft orbiting Earth revealed that the highly energized population of electrons of the inner radiation belt is organized into structured patterns that resemble slanted zebra stripes.

To understand how the magnetic field affects the electrons, one researcher suggested an analogy: if the inner belt electron populations are viewed as a viscous fluid, the observed global oscillations act to slowly stretch and fold that fluid, much like taffy is stretched and folded in a candy store machine.

The same team found that ring currents generated by solar storms can quickly reduce radiation belt particle intensities around the Earth.

Magnetic storms around the Earth caused by solar prominences can change the shape and intensity of the ring currents within a few hours of the onset of the storms.

Radiation-belt electrons are associated with ionized oxygen and hydrogen particles which contribute to the ring current dynamics. Peaks in ionized oxygen, measured in the range of 50 to 180 kilo-electron volts (keV), are associated with storm activity on the side of the Earth facing away from the Sun. The storm-time ring current produces significant distortion in the Earth's magnetic field. The distortion takes the form of electron drift path "islands," or regions of almost zero-magnetic field strength within the otherwise dense magnetic field that surrounds the Earth.

The islands can be thought of as local minima in the Earth's magnetic field. Whether or not the zebra banding effect in the radiation belts has any effect on the dynamics of these local minima, or vice versa, has yet to be determined. However, in the black-hole context it seems likely that the Earth isn't solely responsible for the zebra pattern. The Earth's rotation is the result of torque applied from without. (Refer to Chapter 3 and wormholes.) The torque is a product of four-dimensional spacetime structure. Therefore the Earth doesn't turn the stripes, but rather both the stripes and the Earth are turned by the very structure of space.

Black Hole Ergosphere versus Van Allen Belt

The dynamics of the Van Allen belt may be more understandable by the example of a black hole *ergosphere*. The ergosphere is the area between a black hole's outer redshift horizon, also called the static limit, and the event horizon. The inner boundary of the ergosphere is in essence the event horizon. The redshift horizon is the static limit inside of which the black hole's rotational forces predominate.

As a particle approaches a black hole surrounded by an ergosphere, its orbit around the black hole will have a large "binding energy." Its "most bound" orbit lies on the innermost surface of the ergosphere, closest to the black hole event horizon.

The most general type of black hole, and one that looks something like the Earth surrounded by the Van Allen belts, has angular momentum L, linear momentum *p*, charge *e*, and mass *m*. By virtue of these parameters the black hole has a magnetic field and is rotating. A simulation of gas accreting onto a rotating black hole by John Hawley of Caltech and Larry Smarr of the National Center for Supercomputing Applications is shown below. It is very similiar to the picture of the Van Allen belts shown above. This image was made in 1996. More work by Larry Smarr can be found in Chapter 8.

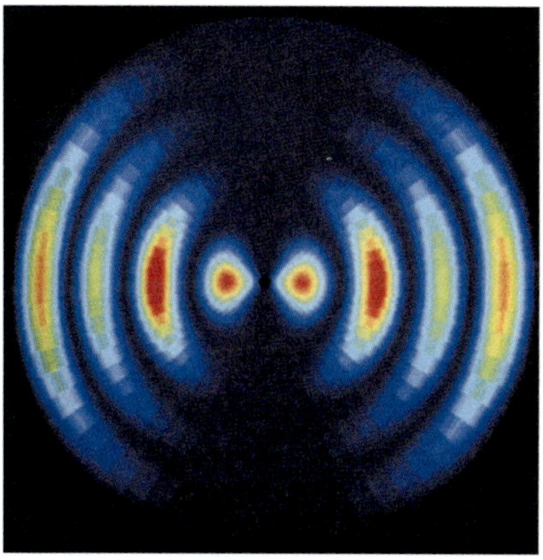

These *accretion rings* can be thought of as the fine structure of the ergosphere. In general, two interactions are possible. First, a particle can dip through the ergosphere and escape with some of the energy and angular momentum of the black hole. Second, a particle that's captured can increase the energy and angular momentum of the black hole. The degree to which energy is added depends on an impact parameter for a captured particle. This in turn depends on how close the particle comes to the event horizon. A head-on impact has a large impact value and thus increases the angular momentum and mass as the particle is captured. A glancing impact will increase these things less.

Part of the black hole's mass is relative, and part is built-in, and the ergosphere can be thought of as the source of the hole's *relative* mass and *relative* angular momentum. A glancing impact will only add angular momentum and relative mass, but the *irreducible mass* will remain the same. Irreducible mass is the mass of the black hole in the absence of rotation. If a particle is captured out of a less favorable orbit, the irreducible mass will increase, because the angular approach adds nothing to the relative mass of the ergosphere through sympathetic motion. The particle just goes right through. If the irreducible mass increases, this also causes the event horizon to expand. (But not the redshift horizon, as we shall see.)

A positive impact parameter results from a particle injected into the system in the same direction as the black hole's rotation. In fact, a particle at rest will automatically be dragged with the direction of rotation and will have a positive energy of motion, and thus will have a small positive impact parameter that goes with the flow, so to speak. This is called favorable or selective accretion.

The true dynamics of accretion at the center of a black hole are still a mystery. Outside, the Sun's quadrupole acts to sweep up matter and dust that swirls towards the Sun along the ecliptic plane. Inside, the dust continues to travel along the ecliptic, but is swept through the spacetime of the Moon and into the Earth's dipole magnetic field, where it's attracted by the Earth's gravity and falls to the surface. The Earth is known to accrete 14.3 million metric tons of cosmic dust per year. The rate of accretion for interplanetary dust as opposed to cosmic dust is 0.62 million tons per year. About 3.65 million tons of meteoric dust also reaches Earth annually.

In a favorable accretion model the particles are aligned with the debris from the collapse of a rotating star, moving coherently and falling into the black hole at a glancing angle, not head on. They are therefore synchronous. Particles injected at a glancing angle increase the star's angular momentum, but only up to a point. Rotational energy can only account for less than 29 percent of the total energy of the black hole. When this limit is reached, no further speedup is possible.

In a *random* accretion model the infalling matter approaches at all angles to the rotating ergosphere, leading to a gradual decrease of the relative angular momentum and relative mass of the system because there's no synchronicity going on and thus no conservation of the relative mass. This is an example of entropy and is an irreversible interaction (or as the mathematicians say, an *irreversible transformation*). However, whether random or selective, the irreducible mass, i.e. the mass of the black hole in the absence of rotation, can never decrease. It can only increase or stay the same. Thus the ergosphere is a homeostatic mechanism, to use a medical term, for the life of the star. It keeps things in balance.

The ergosphere arises from the presence of angular momentum. This describes a Kerr geometry. A degenerate case of the Kerr geometry is the Schwarzchild geometry in which the event horizon and static limit coincide. Otherwise the two surfaces are separated everywhere except at the poles. The outer limit of the ergosphere is a surface of infinite redshift, which is the same as saying that particles are captured at the speed of light. In fact particles in the Van Allen belt are "relativistic," meaning that particles approach the speed of light but never quite get there. The particles of a black hole jet are similarly observed to be relativistic.

The ergosphere is only one part of a bigger mathematical system. Though 29 percent of angular momentum or rotational energy is the maximum potential of the ergosphere, a built-in angular momentum comes from the *horizon framework*, as will be discussed below. The framework generates a base mass and angular momentum. The ergosphere adds to or subtracts from this base energy, something like the spinning wheel of a gyroscope, with the frame of the gyroscope being analogous to the "framework".

An ergosphere has a dragging effect on space. The effect is very small for the Earth, but supposedly dragging would cause a perfectly spinning gyroscope orbiting above the Earth to precess. This effect might explain how the Van Allen belts can rotate with the Earth, and at the same time precess West by 0.3 to 0.5 degrees per year. The SAA drift would thereby be a relativistic physical effect.

The increasing size of the SAA is another story altogether. Ridley, Gamow, Schrodinger, and originally George Darwin, the son of Charles Darwin, pointed out that the Moon is constantly receding from the Earth. Its distance from the Earth increases by about five inches per year. The Earth rotates within two tidal bulges caused by the orbit of the Moon. As the Moon recedes, the rotation rate of the Earth between the tidal bulges will simultaneously decrease, causing an increase in the length of the day.

This effect is explained by the law of the conservation of angular momentum. This law tells us that the reduction of the Earth's angular momentum must be balanced by an increase in the angular momentum of the Moon in its orbit around the Earth. The total linear momentum and the total angular momentum of the combined system remains constant.

Schrodinger points out that in the motion of the celestial bodies, irreversible frictional and thermal influences are present. This directly relates to the friction of accretion, or to the "random" model of accretion, by which the Earth would lose relative angular momentum. The rotation of the Earth is slowly diminished by tidal friction because the Earth is not a completely rigidly rotating sphere.

The fact that the angular momentum of the entire system is conserved is good news, but the manner in which it is conserved requires some thought. In a black hole context, the system is concentric and so the Moon is going nowhere.

In the Einstein field equations for a black hole, the best fit for the observed lunar and solar periods is found for a compensating inner system of horizons which represents an evolved state of the black hole system relative to "canonical" values. The canonical system has slightly less mass and angular momentum than the inner system. Thus the current "place" of the solar system in the black hole evolution is associated with a higher-mass state than the canonical state.

Given this finding, the Moon is associated with the inner horizon of the inner system. Likewise, the Sun's best fit is associated with the event horizon. The equations indicate that as the system gains mass and angular momentum, the turning point or inner horizon (that reflects the Moon) shrinks, or approaches the Earth; while the event horizon (that reflects the Sun) expands, or recedes from the Earth. Meanwhile, the outer horizon also shrinks as the system gains mass.

Thinking optically, with the horizons as reflecting lenses, then as the *inner horizon* shrinks, it will magnify the image of the Moon less, so that the Moon will appear to get smaller and so to recede. Likewise, as the *event horizon* expands, the Sun will appear to get bigger over millions of year of evolution. Meanwhile, the stars reflected on the *outer horizon* will, like the image of the Moon, appear to recede over time. The outer horizon, called the redshift horizon, already traps relativistic particles which show a time dilation effect to observing spectrographs, and so presents a redshift that would make the stars appear to be receding faster than they actually are, if at all. But the truth of the matter is that the outer horizon is physically approaching the Earth over time.

At least that's the truth of the matter according to mathematical theory. The real truth is that the horizon equations are incredibly fragile, meaning that the choice of variable values is critical. Also, this behavior is based on the major assumption that the canonical solution represents a median state of the star's evolution. Also, past certain limits the equations are ill-defined. In reality there may be equilibrium processes going on in the body of the star that compensate for these dynamics. Nature is surprising. Math is predictable.

Playing the math game some insight can be gained. We set up a horizon framework where, first, the Sun is associated with the event horizon, as with the Einstein solution. Second, the reflection of Jupiter is associated with the outer or redshift horizon. The first question is whether the expanding event horizon produces a slower solar period or a faster one. Logically the action of the Sun should be reversed for the contracting outer horizon associated with Jupiter.

Using the Jupiter opposition period of 398.89 days as a check, and

365 days for the Sun, and as well 4334 days for the Jupiter sidereal period, all observed values, a basic Euler convergence can be run on the following bracket values to get the following results:

Sun:	366	365	364	(event)
Jupiter:	4330	4334	4338	(outer/inner)
Result:	399.79	398.56	397.34	

These numbers show that to get the "pure solar" value of 400 days (approximated in the left column), the expansion of the event horizon is perceived as a lengthening of the Earth year – in accordance with actual observations of the sidereal year; while the contraction of the outer horizon is perceived as a reduction of the Jupiter orbital period. In general the numbers show that, all things being equal, as the framework gains mass, the Jupiter opposition period gets longer.

Note that the bracket values used in the table above are large departures from the current values in the middle column; the changes are in reality much much smaller.

Theoretically, if an astronaut was able to dive towards the core of the star, the star's evolution could be forced so that the event horizon expands, and the outer horizon contracts, showing a "core solar" opposition period of 400 days. But at the Earth distance, we get 398.89 days. If the astronaut was to travel out of the star, the opposition period would decrease to 397.34 days. Thus the general rule of "looking out of a black hole, time speeds up" isn't quite accurate. It's more like, escaping a black hole, time speeds up; and falling into a black hole, time slows.

The Firewall

In 1929 Edwin Hubble observed that the universe is expanding. Not only that, but that stars farther away are moving faster away. Given the Einstein equations discussed in the previous section, how can the proposed scenario be reconciled with Hubble's observations? If the stars are reflected off of the outer horizon, where must they be?

Likewise, the Moon may be reflected off of an inner horizon. If the Moon is inside of the Earth, then it's possible that the stars are too. We see the sparkling expanse of a sky which is, actually, a reflection of the real stars on "the other side of the Moon". But the language here is very deceptive! When a person normally thinks about the "other side of the Moon" they think of, simply, the space beyond the Moon, and assume that the whole thing is "outer space".

But the truth may be more subtle. If we see the reflection of the stars, then the real stars are indeed "on the other side of the Moon" – given the understanding of the Moon being interior to the Earth. The Earth would be a very protected and secluded environment from an outsider's viewpoint. One can think of the Earth as being behind a firewall of dark matter.

In 1952 Truman Bethurum wrote a book about his experience with "aliens" – human-looking aliens, but probably people from nearby military installations in Nevada. Bethurum distilled their knowledge in the guise of a female space alien, to make the truths more palatable. But one comment makes total sense with the subtle understanding outlined above.

Bethurum wrote: "I told her that if I remembered correctly the first time I had met her she had told me that the name of her planet was Clarion. I asked her if she thought that we on Earth might know it by some other name, such as Mars or Jupiter. She smiled and assured me that such was not the case, couldn't be, since her planet would be entirely invisible from Earth, since it was on the other side of the Moon. Being on the other side of the Moon, Clarion was hardly likely to be visible to us from any point on Earth."

Obviously the Moon orbits in such a way that the entire universe is visible and the Moon blocks none of it permanently. In the normal sense. But the subtle interpretation is that Clarion, like all the stars and planets, is on the other side of the wormhole that makes up the Earth-Moon system, and so is "on the other side of the Moon". It's probable that the language is slightly corrupt, that what Bethurum was being told and what he reported are slightly out of step with each other – because planets as well as stars are visible at least as a reflection on the Earth side. And yet, forgiving the error of reportage, there is the possibility of an immense truth in this meeting of the minds!

Getting back to the actual source of these ideas, being the military in and around Nevada, one can switch things around so that this race from the other side of the Moon who found Earth exceptionally difficult to reach, reflects the reality of scientists on Earth having exceptional difficulty in reaching the Moon. However, there's no record of an American attempt to reach the Moon until Pioneer 4 in 1959, which passed within 36,000 miles of the lunar surface.

Local Contamination of the Microwave Background

In 2005 the measurements of the cosmic microwave background (CMB) were found to be either contaminated by the solar system or some other foreground source. The contamination was found in the original Cosmic Background Explorer (COBE) data, and confirmed in the Wilkinson Microwave Anisotropy Probe (WMAP) data.

Starkman and Schwarz described the contamination as a lack of large-angle correlations. The sphere of space that surrounds the Earth, and which is shown in the Planck images, can be analyzed in different modes. In other words, the sphere can be divided into halves, called a dipole, or quadrants, called a quadrupole. A dipole has three modes corresponding to the three axes of space: forward and backward, up and down, and left and right. A quadrupole, as shown below, has five modes because unlike the dipole two extra modes are possible mirrored on either side of the equator. Two quadrupole modes are shown below. On the left are straight sectors, and on the right are how they look when mapped to CMB coordinates irrespective of any anomalies.

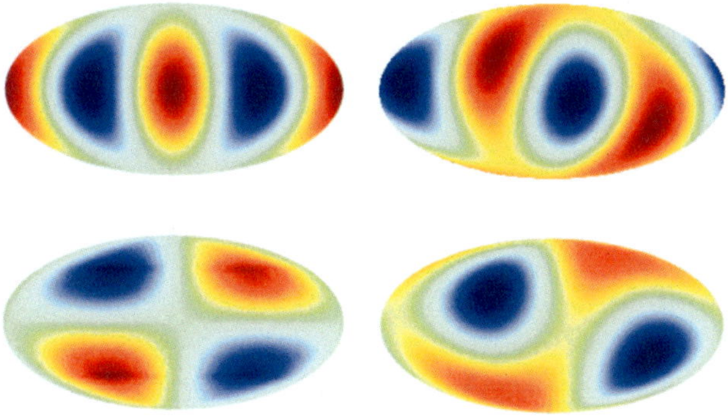

Scientists expect the cosmic background to follow a curve described by an angular correlation function which relates points in the sky separated by a given angle. However, the theory doesn't match the observations. Above an angle of about 60 degrees, the COBE and WMAP data zeroes out and departs from the theoretical curve. This is called a lack of large-angle correlations.

The worst error is associated with quadrupole and octopole divisions of the sphere. The Planck mission, though looking at the galaxy with a wider frequency range than previous missions, offered no new data on the problem. The IBEX findings may have some bearing on the problem with the discovery of an energetic band at the heliospheric boundary corresponding to the solar system's ecliptic plane.

Scientists believe the culprit is contamination by the solar system. Even though the COBE and WMAP mathematicians took the solar system into account and filtered it out mathematically, what's left over still acts as a significant "foreground source" that causes flattening of the expected curve.

Now, if the Earth is surrounded by event horizons upon which the stars are reflected, the contamination could come from the energetic fields associated with those horizons, much closer to home than other high-energy constructs in the local space that extends beyond the solar system. Starkman goes a step further and postulates that the universe as a whole may not be in the shape of a sphere but may wrap around on itself like a donut or torus. This would produce a distinctive modification of the data, though he doesn't say whether a toroidal structure can be directly attributed to the contamination. This hypothesis is difficult to verify because any toroidal structure may be just over the inflationary horizon, and thus difficult to see.

But what does this mean? Michio Kaku explains: "Inflation theory, proposed by Alan Guth of MIT, states that there was a much more explosive expansion of the universe at the initial instant of Creation than the usual Big Bang scenario; it holds that the visible universe we see with our telescopes is only the tiniest part of a much bigger universe whose boundaries lie beyond our visible horizon."

And so a new interpretation is that, if indeed the connection between the Earth and Moon is some form of a, in mathematical terms, cobordism, then the part of the universe we see from the Earth is only a small part of the extended universe as seen from the far side of the Moon. But you have to travel through the Earth-Moon wormhole to get to the Moon and thus to see the "open side" of the universe. Behind the firewall, the Earth is very much embedded in a fixed environment.

The geophysical view of the Earth and Moon indirectly supports this idea, in that the order of solar burning seems to be backwards if Earth is the "outside". Earth's core is made of iron, which is the last stage of solar burning. If Earth is on the inside relative to the "outside universe", then it makes more sense. The Sun, reflected from within, is fixed in the future and burns hydrogen. The Sun's helium-rich core grew hot due to gravitational compression and ignited. The ash of helium is carbon and oxygen. Carbon and oxygen at the star's core then started to burn. Carbon burns first, and then oxygen. The ash of oxygen burning is silicon, which makes up most of the refractory surface of the Moon. The ash of silicon burning is iron, which, farther in, makes up the Earth's core. And most marvelous of all, from some reaction or process that no one understands, though iron leaves no ash behind and is the end of the line of solar burning, the hot iron core somehow interacts with and is completely compatible with … water. The innermost shell of the burning Sun, seen from this inside-out perspective, is the water environment of the Earth.

Chapter 7: Climate Change

SPA and Eltanin

The South Pole-Aitken (SPA) basin on the far side of the Moon is thought to be the largest "impact" crater in the solar system, but since the SPA is associated with the lunar south magnetic pole, the likelihood that it was created by an impact seems slight. The SPA is centered on lunar latitude 55 degrees S, longitude 180 (the anti-meridian).

The SPA finds an analogue on Earth with the Eltanin "impact" crater, which scientists are still actively researching. The Eltanin impact deposit is characterized by large concentrations of melted and unmelted meteoritic material distributed across a large area of the ocean floor. The Freeden seamounts rise about 2700 meters above the surrounding ocean bottom (the "abyssal plain") in the area affected by the Eltanin impact.

Scientists estimate that coarse ejecta from the impact covers more than a million square kilometers of ocean floor, equivalent to a crater with a diameter of 1128 kilometers. As yet, scientists have not explored the areas to the south and west of the seamount area, which is located at 57.5 latitude S, and 90.5 longitude W. Again it seems unlikely that such a huge ejecta area can be attributable to a meteor, but scientists have already postulated tsunami and air fallout scenarios for the impact event. It seems more likely that this is all related to mantle material cycling to the surface, in the same way that craters evolve on the Moon.

Scientists think that the Eltanin impact took place about 2.5 million years ago, however the SPA on the Moon is so old that it has been covered over by more recent crater activity, and was only discovered with precision instruments on the Clementine and Lunar Prospector missions. It was also initially obscure because of its vast size. The SPA basin is 2500 km in diameter and covers an area of nearly 5 million square kilometers. It's thought to be the oldest feature on the

Moon, excepting perhaps Procellarum.

Lunar scientists, like their terrestrial counterparts, are also focussed on meteorite impact scenarios. They think that huge quantities of ejecta were deposited outside of the SPA basin, thickest near the rim. In theory the ejecta deposited nearest the basin rim contains material from the Moon's lower crust and even from the mantle, due to the estimated force of the "impact". Scientists know for sure that the SPA interior has a 7 to 10 percent enrichment of FeO (iron oxide) relative to the surrounding highlands.

The Eltanin crater covers a deep-ocean basin 5 kilometers deep, and the central portion of the impact region is the most meteorite-rich locality known on Earth. For comparison, there is a 13 kilometer difference in elevation between the interior of the SPA basin and the surrounding highlands. This just highlights the fact that there will be differences between Earth features and Moon features. Moon features are bigger not because the Moon is bigger than the Earth, but because the Moon is more exposed to the Sun. Analogues like Eltanin and the SPA won't be exact replicas, but there will be a considerable overlap in features. The course of evolution on the two planets is similar but not identical.

The spherules examined from the Eltanin cores consist of crystallized or shock-melted pyroxenes and olivines, and a siginificant fraction of spherules are glass. Olivines are part of the Earth's mantle, and melting or differentiation in terrestrial olivine shows the same Lithium composition as differentiated basaltic mantle material on the Moon.

Scientists believe that the starting material for the Eltanin asteroid was basalt. All of the Eltanin spherules are enriched in TiO2 (titanium dioxide). The interior of the SPA also exhibits enhanced TiO2 levels, which drop off with increasing distance from the center. One could attribute the similarity between craters to the mechanics of an asteroid or meteor impact, which is after all thought to be the cause of both anomalies. But a much more conservative and real explanation is that, as already discovered by scientists, the Earth and Moon mantle reservoirs have essentially the same composition.

The disposition of the Earth's geomagnetic field can be described by the conceptual diagram shown below.

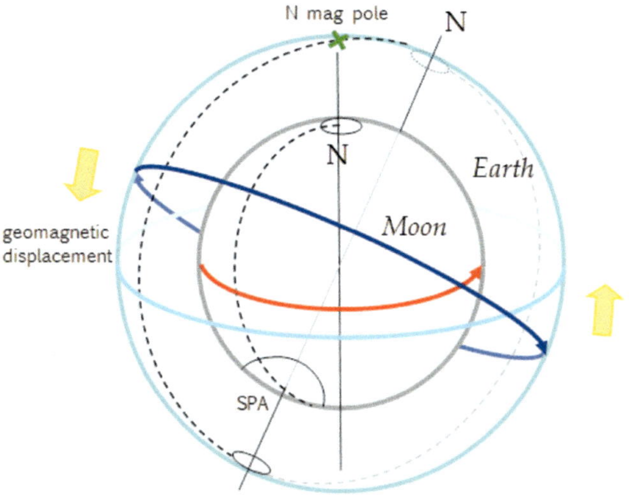

This isn't a physical depiction of the Earth and the Moon, since obviously astronauts don't drill 2000 miles into the Earth's crust and mantle to get to the Moon. Rather, it's a conceptual model showing how the Earth and the Moon, when combined in a concentric arrangement, produce a physical consequence, being the magnetic positionings on either globe.

It's not necessary for the Moon to be interior to the Earth: the arrangement could be reversed, with the Earth interior to the Moon, and the consequences would be the same.

Since everything is relative, one can dub this a "relativity diagram". The sizes and angles are exaggerated for ease of viewing. The Earth's north pole is just over the horizon at the top of the globe, and the geomagnetic meridians (dashed lines) of the Earth and the Moon are supposed to be aligned. The meridians on the visible halves of the globes are at 102 W longitude on Earth, and 180 longitude on the Moon. For simplicity, the greatest geomagnetic displacement on the

Earth globe is shown at the indicated longitudes; in reality the greatest displacement is at 69 W longitude and 111 E longitude. That displacement is about 12 degrees south at the western meridian, and 12 degrees north at the eastern, compared to the Earth's *geographic equator*.

The tilt of the Earth is 23.5 degrees, exaggerated in the diagram. The Earth is tilted relative to the ecliptic plane of the Sun, and the Moon's equator basically aligns (within a few degrees) with that ecliptic. The Moon is tilted, in the conceptual diagram, 22 degrees relative to the Earth's poles. But the geomagnetic tilt is less than the Earth's axial tilt: 12 versus 22 degrees.

The latitude of the SPA region is 55 south; and likewise the latitude of the Eltanin feature on Earth is 57.5 south. The simplest interpretation is that in the past, the poles of the Earth and Moon were aligned. How far back? The "migration" probably happened after plate tectonics got started on Earth some 3.8 billion years ago. It may have lasted over a lengthy period of, say, 800 million years. As the mantle became hotter, the hot material became less dense, and so opened the way for shifts in weight distribution, especially if the Earth's core formation was "accentuated" or zonal rather than uniform around the entire circumference of the Earth. (Recall from Chapter 2 that convection probably began in a ring around the Earth's equator, which, significantly for this sequence of facts, was then aligned with the orbit of the Moon.)

Today we know that the Moon is tilted 22 degrees relative to the Earth. However, the geomagnetic meridian is tilted halfway between the Moon's embedded equator, and the Earth's equator: about 12 degrees. Remember that the geomagnetic field is *the Earth's* geomagnetic field. It emanates from within the Earth, and given the Moon tilt of 22 degrees, the relative tilt of the geomagnetic great circle may simply be a result of the distance the electromagnetic field has to travel through the Earth's mantle. The mantle acts to deflect the field slightly.

On the other hand, as shown by the image below, the high atmosphere is electrically charged along the solar ecliptic, at a steeper incline than the geomagnetic parallels. The tilt here corresponds to the Earth's tilt of 23.5 degrees.

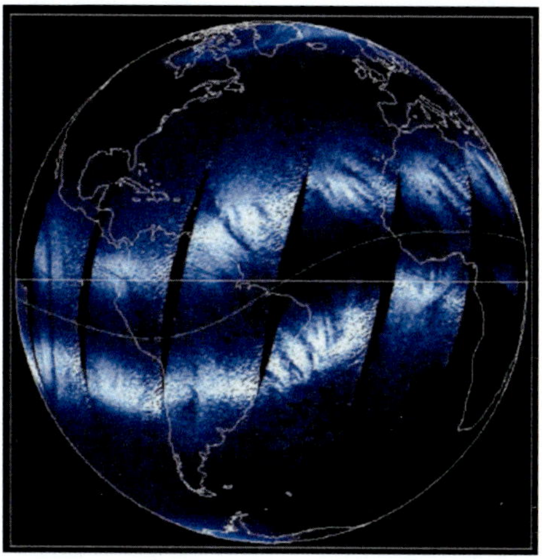

These "dayside maxima" correspond to the Earth's face in full sunlight, and so the Sun's electromagnetism is thought to charge the atmosphere and produce electron density peaks in bands 30 degrees to the north and south along ecliptic lines of latitude. These peaks are more steeply inclined than the Earth's geomagnetic field, and this energetic radiation is more powerful than the radiation associated with, for example, the aurorae, which are seen polewards of +/- 45 degrees geomagnetic latitude. The aurorae have no effect on the Earth's climate, but the electron density peaks do.

The Hot Earth

In 2011 NASA launched the Polar Orbiting Suomi satellite. It orbits 500 miles high and circles the planet 14 times a day. The CERES instrument onboard this satellite measures energy in the ultraviolet and infrared (heat) bands. CERES stands for Clouds and Earth's Radiant Energy System.

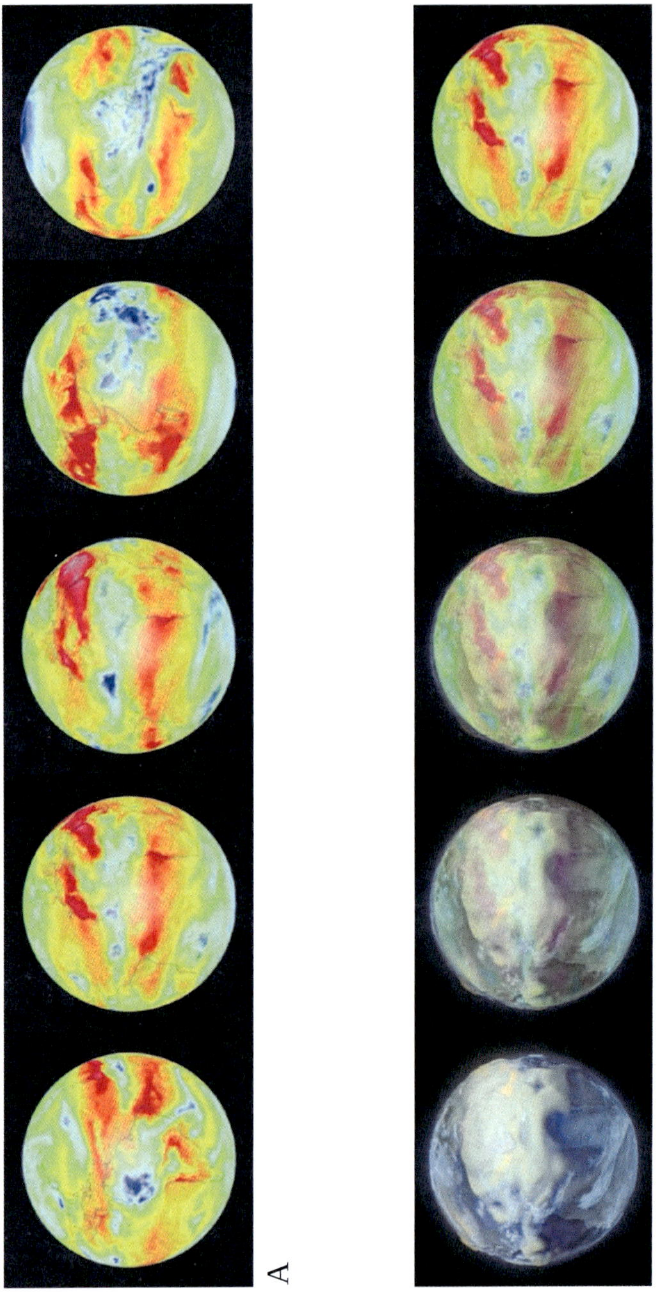

A

B

The graphic A (facing page) is from an animation of the Earth's radiant energy produced from the CERES sensor data. It shows heat absorption and reflection in the atmosphere. Blue denotes the reflection of heat, red is heat absorption, and green is steady state. The data clearly shows two bands, north and south, corresponding to the ionization of electrons in the F2 layer of the atmosphere, some 186 miles in altitude.

Clearly there's a correspondence between heat absorption and the electron density peaks produced by the Earth's magnetic field. But this new data also shows that only the northern band follows the ecliptic field lines faithfully; the southern band is less defined. The predominance of the Indian Ocean in the southern hemisphere may have something to do with the southern band's flatness – it seems to follow the geographic lines of latitude rather than the geomagnetic lines.

For comparison, the second frame of the series above is superimposed with a similar view of the Earth in terms of nitrate production (the leftmost frame of B, again produced by NASA). Nitrate is formed in the atmosphere from lightning, which breaks up nitrogen molecules which then recombine with oxygen. The nitrate levels are shown in shades of yellow in graphic B, the brightest yellow being the most nitrate production.

Though the nitrate data and the CERES data were collected at different times, and though the shape of the distribution changes from day to day, nonetheless it follows a definite line, in that nitrate is produced in the places where heat is not absorbed – that is, in the central "trough" between the heat bands, north and south. This central space corresponds to the Earth's geomagnetic equator. We can reduce everything to the geomagnetic equator since it's mapped to the same place on Earth even as the Earth rotates.

Since heat absorption translates to dry regions on Earth's surface, nitrates and lightning seem to like fertile regions along the geomagnetic equator, and favor to a lesser degree regions outside of the heat bands at higher geomagnetic latitudes. It's hard to say which is the cause and which the effect: nitrates or fertility, since one is very much tied to the other. Nitrates fertilize the soil through rainfall, but without moisture there could be no lightning and thus no nitrates.

Ocean salinity also has two bands, north and south, that circle the globe. The salinity is measured as a ratio of the electrical conductivity of a seawater sample and the conductivity of potassium chloride, which acts as the control. This ratio is called a “practical salinity unit” or “psu”. Of course conductivity is a major part of the magnetic field of the Earth, but the actual dynamic of what causes increased salinity is more related to heat. According to the NOAA, processes which increase a body of water’s salinity include evaporation and freezing. One can posit that where the conductivity and therefore the heat of the ocean surface is greatest, there’s more evaporation and thus higher salinity. By contrast, dilution of seawater is caused by river runoff, ice melting, and significantly for the two bands, precipitation. As already mentioned above, the middle area between the bands is a fertile region of rainfall, and, in agreement with this, an area of decreased salinity.

As mentioned in Chapter 1, global warming is known to be causing the sea levels to rise. However, the actual cause of most of this sea level rise is probably not the melting of the ice caps, but the heating of the oceans in precisely the bands discussed above. The maximum sea level rise along the double bands is roughly 15 centimeters (6 inches) over a period of roughly 25 years, a measurable increase and therefore of concern.

Water expands as it gets hotter, so sea level data can help scientists understand how currents are changing and how the ocean is storing heat. Total sea level change between 1992 and 2019, based on data collected from the TOPEX/Poseidon, Jason-1, Jason-2, and Jason-3 satellites, is about 4 inches on average. Given that the Earth is intimately tied up with the cycles of the Sun, it’s not surprising that solar cycles influence things like water levels on Earth. Everything is inter-related.

Part of global warming is *desertification*, and it’s not surprising that the world's deserts follow the great circles of the geomagnetic bands, north and south. A stunning map published by NASA's SeaWiFS Project clearly shows a direct correspondence between the geomagnetic bands and landmass areas of minimum vegetation. (See this book's back cover.) Earth's magnetic field, whether related to an interior Moon or not, creates the blueprint for where life can exist on Earth. Measurements by satellites provided estimates of terrestrial vegetation and sea-surface chlorophyll during a program that mapped the global primary biosphere from 1997 to 2000. There's nothing mankind can do at a stellar level to change the reality of desertification; but where the Earth is habitable, we are duty-bound to respect the processes that support life. We have a responsibility to cherish and protect our special planet.

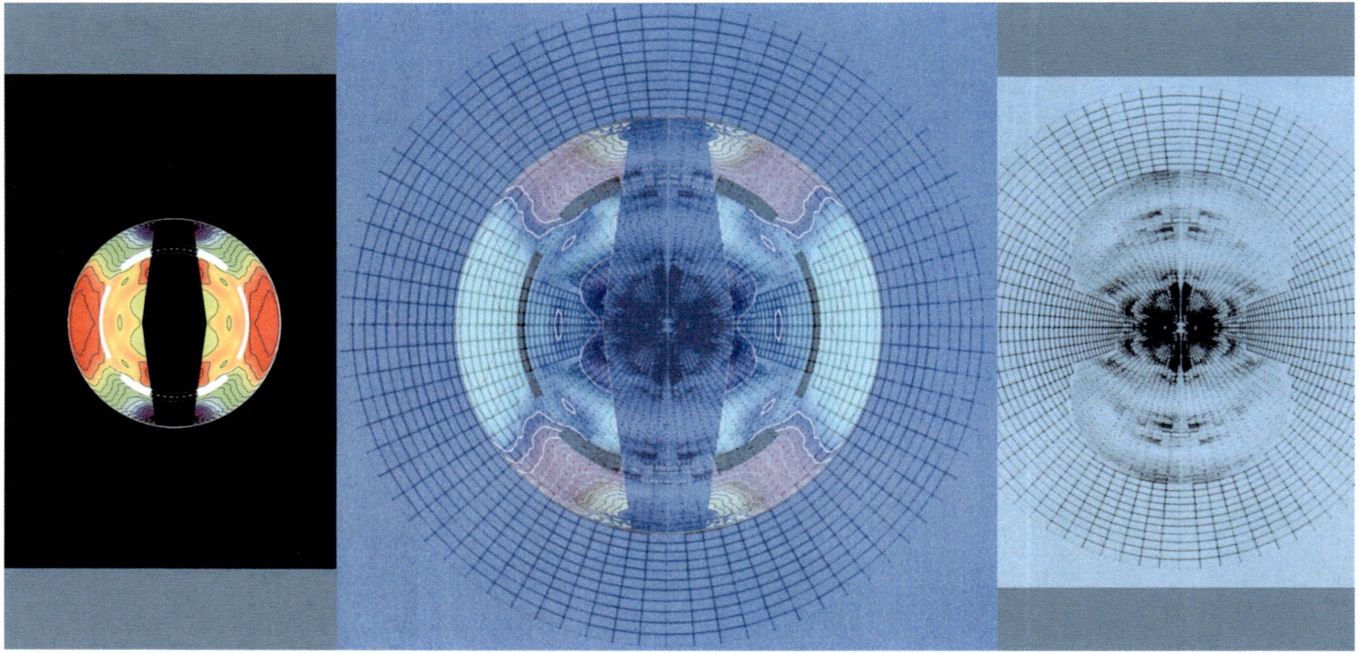

underlay

overlay

Chapter 8: The Sun

The Underlay

The world is full of wonder, and of course one of the most wonderful things of all is the Sun. The image on the facing page is a composite image made from an underlay image (left) matched with an overlay image (right). The underlay (left) was published in 2007, and shows the inferred rotation rate of the Sun based on 144 days of observation by the MDI (Michelson Doppler Imager) instrument on board the SOHO (Solar and Heliospheric Observatory) spacecraft. The MDI project was a joint effort of the Hansen Experimental Physics Lab at Stanford and the Lockheed-Martin Advanced Technology Center. The MDI project started observations in 1996 and came to an end in 2011.

The Solar Oscillations Investigation (SOI) that employed the MDI was an international collaboration to study the interior structure and dynamics of the Sun. The data for the image was collected in 1998. The Sun's convection zone is shown as the outer region, and the Sun's core as the inner. They're separated by the *tachocline*, a thin region shown in white which comprises only four percent of the solar radius. The convection zone contains only two percent of the mass of the Sun. Almost all of the Sun's mass is locked up at the core.

Very sensitive instruments are required to measure the core of the Sun, which is why the center column is missing in the underlay. In order to map differential rotation, a set of resonant modes of the Sun are observed as frequencies (for radial structure) and as frequency splittings (for rotation). Various wave-modes produce an average oscillation of 430 nHz at the upper transition zone of the Sun's core. Inside the tachocline, the entire core rotates at the same speed at all latitudes. Outside the tachocline, in the convection zone, the rotation is a function of latitude. The fastest rotation is at the equator, with an oscillation of 460 nHz, and it decreases with higher latitude towards the poles, to about 330 nHz. This translates to 27 days for the core rotation period; 25 days for equatorial rotation; and 35 days for the polar rotation period.

The Overlay

The overlay (plotted image, right) was published in the popular press in 1982, and is a joint collaboration between John Wheeler at Princeton University, his unnamed graduate student, and Professors Larry Smarr and Ken Eppley who were working at Lawrence Livermore Labs, who wrote the code that ran on the facility's supercomputers. The plot is the output of a computer simulation of the gravity waves resulting from the collision of two black holes.

One must set the way-back machine to 1978 or 1979, when many long years of work were coming to fruition in numerical computing. The overlay may look like a piece of artwork, but in fact every pixel was computed using very complicated equations. Since simulations based on theory are the only way we'll probably ever see the interior of the Sun, it's worth going into.

The Simulation

The simulation takes place in a "frame" using the ADM 3+1 spacetime formalism, where gravity terms are split off into a 3-metric (with shift vectors factored in), and the "+1" part is a lapse function. Radial coordinates *r* and time coordinate *t* are selected in this frame. The 3-metric is evolved to simulate the infall of the two colliding black holes. Each "frame" of the evolution *T* is further split into time slices *t*. Each slice is a surface upon which the lapse function works normal to the surface; and the shift vector operates *in* the surface. Given a frame, gravitational radiation information can be read out of a spacetime. The lapse function and shift vector fix a computational "gauge" or "gauge field," which is tied up with symmetries in the gravitational field, e.g. in the geometry. For example, a gauge symmetry operation such as rotation on a plane can be applied as 45 degrees for one point, 60 degrees for another, and 90 degrees at another, and despite the nonuniform application of the symmetry, the equations of motion which control the dynamic evolution of the fields don't change, and neither do the overall physics. Everything looks the same. Symmetry operations include rotation, translation, and reflection.

The frame globally covers the spacetime region being simulated, with the two black holes providing a “strong field.” Symmetry simplifies the computation: the simulation is axially symmetric, or *axisymmetric* for short. This means only one quadrant needs to be computed, and it’s flipped across the vertical axis, and then the whole thing across the horizontal axis. (Look at the plot shown on the right to see this.) The 3+1 metric, as opposed to the full Einstein 4-metric, leads to the least number of terms in the field equations. This is because a clean geometric split between kinematic variables (lapse and shift) and dynamic variables (the Einstein 3-metric) occurs in the 3+1 approach. If one can trade dynamic variables for kinematic ones, then the resulting Einstein equations are much simpler and easier to compute.

Black holes both absorb and radiate energy, the same as all other stars, including our Sun. Gravitational energy flux is computed by splitting spacetime into *spacelike* hypersurfaces and *null* hypersurfaces. A null hypersurface has normal vectors with zero length at every point with respect to the local metric tensor. In the context of the simulation, a null hypersurface describes an event horizon. In theory as time comes closer to the event horizon, the degree of curvature becomes ever larger, to the point where Einstein’s equations break down. One can “look inside” an event horizon, but eventually the curvature becomes singular. A spacelike hypersurface is simply a slice through four-dimensional spacetime in the form of a two-dimensional sheet of space. Since spacetime around the black hole is curved, this slice, though two-dimensional, will also be curved. The 2-sphere at time t surrounding the radiating system (indicated by the radial grid in the plot) will overlap one type of hypersurface or the other. The simulation also takes into account flat spacetime far from the black hole that surrounds the extreme curvature of spacetime around the black hole. The black hole is smoothly connected to the flat spacetime of the rest of the universe. This “asymptotically flat” frame of reference prevents fundamental properties of the black hole like mass, charge, and spin, from going out of control. Flat spacetime is like an “anchor” on the variables.

The closer the computation comes to the black-hole event horizon, the less stable the equations become. The way Smarr worked around this was to slow down time, and this in turn was accomplished with "maximal surfaces." A maximal surface is a spacelike hypersurface with a vanishing mean extrinsic curvature everywhere. The selection of a time coordinate *t* close to the singularity is guided by extrinsic curvature. Extrinsic curvature describes the manner in which the spacelike hypersurface is embedded in time. It's part of the evolution equation and, if one thinks of the classic "kitchen sink" diagram with water going down the drain, can be viewed as the velocity of the dynamical 3-metric. It has no relation to lapse and shift but rather measures curvature intrinsic to the time slice. Lapse and shift make reference to the spacetime coordinates in the neighborhood of the null hypersurface. The shift vector determines how the spatial coordinates on one time slice are pulled to a neighbor slice, but in the vicinity of a null hypersurface things can get tricky. The simulation therefore proceeds from the null hypersurface outward. This in turn leads to the selection of "initial conditions."

A maximal surface aids in setting the initial values of the gravitational variables on the initial slice. A "minimal distortion" shift vector is employed to minimize the change in the 3-metric with respect to the spacelike hypersurfaces, and to prevent the vector field from falling inside of the null hypersurface. This is related to the trace of the extrinsic curvature tensor along the spacelike hypersurface. Since the evolution from t_0 to $(t_0 + t)$ involves two hypersurfaces, there are two traces or curves, each in its own plane. The trace can be moved in such a way that the moving frames of the two curves coincide. Another name for this is "congruence."

Smarr notes that congruence can be used as a control. For two curves or surfaces in motion, if the surfaces are images of each other, then they are congruent. An *image* is a constant relation between the lengths or surface areas of the curves or surfaces respectively. According to Smarr, spatial coordinate congruence can be selected in such a way to control the amount of "coordinate shear" that appears in the 3-metric at some given time *t*. He notes that gravitational radiation is a propagating transverse shear mode, so the ability to separate coordinate shear from the physical shear of gravity is very important, because otherwise the results would be slightly off. The minimal distortion vector is actually a shear equation that minimizes the square of time "rate of change" of the conformal metric.

In short, maximal surfaces, extrinsic curvature, and minimal-distortion shear are all related and yield robust initial conditions on the metric. That said, it turns out that maximal surfaces are solved with elliptic equations. Solving these nonlinear elliptical equations is the foundation of the gravity computation, and is very expensive in terms of processing time, which is why Smarr and Eppley resorted to running the simulation on a Lawrence Livermore Labs supercomputer.

Stepping Back

The fundamental relationship that makes the simulation possible is the link between Einstein's field equations and Riemannian geometry. The measurement of individual tangent vectors describing a "translation" can change as the computation moves from point to point on the spherical manifold. To handle this variability, Riemann introduced a construct called a *metric tensor* (or curvature tensor) that provides a way to compute the length of a tangent vector at each point. In two dimensions the metric tensor is a two-by-two matrix; in *n* dimensions the tensor is an *n*-by-*n* matrix. The metric allows the length of any curve in a manifold of arbitrary dimension to be measured. As well, the surface area of the space can be measured, and a surface is not limited to two dimensions.

Einstein recognized that spacetime could be described by Riemann's geometry. Einstein's theory of general relativity unifies space and time with mass and gravity. Without curved geometry in higher dimensions Einstein's field equations wouldn't exist. The Riemann metric tensor describes the curvature of spacetime as well as the gravitational field because gravity, in essence, is geometry. The metric tensor is a collection of ten numbers defined at each point in space.

Those ten numbers can be arranged as a 4-by-4 matrix, with six redundancies (being each member with column and row indices *i* and *j* that finds a duplicate in another member with indices *j* and *i*). It's a generalization of Faraday's field concept, not of Maxwell's electromagnetism. In broad terms, the Einstein equations are at the core of the algorithm for evolving the initial state of the gravity simulation at time t_0 to some other state at time $t + t_0$. The first step of the evolution can be viewed as a map $U(t)$ which takes as input $S(t_0)$, the state at t_0, and produces $S(t + t_0)$, that is: $S(t + t_0) = U(t)S(t_0)$.

Before Smarr and Eppley, physicists investigated equations to simulate small perturbations of special solutions such as the Kerr black hole. Unfortunately, such simplifications prevented the study of realistic models for sources of gravitational radiation, such as non-spherical stellar collapse or colliding black holes. This restriction was completely removed in the Smarr-Eppley work. The idea of using sophisticated computer programs to solve the full Einstein equations originated with Bryce DeWitt, whose preliminary work in 1966-74 laid the groundwork for the final assault by Smarr and Eppley. The collision problem is a complete calculation of an asymptotically flat, highly non-spherical, radiating solution of the full nonlinear Einstein equations.

Black holes were found to be more stable and less wild than previously imagined. Part of the reason for the low-energy results is that the collision is "head on," which means that one black hole travels down the positive x-axis, and by symmetry the other travels up the negative x-axis, so that the collision is a "radial infall" rather than a "circular in-spiral." The energy involved in the in-spiral is about a hundred times greater, but the simulation for an in-spiral is more complicated because the symmetry is only two-way instead of four-way as with the radial infall setup.

Time symmetry gives gravity waves an "imploding-exploding" character. The solution contains both incoming and outgoing waves and produces more than one pulse of radiation. Time symmetry means that the evolution can be run in forward and reverse; when time is reversed the velocity vectors and the direction of the gravity field are reversed and the Einstein equations are still perfectly valid. Time symmetry is required because the simulation doesn't model a single black hole but a *collision* of two black holes, which means that one black hole impinges on the other and vice versa due to symmetry.

Each black hole generates gravity waves that travel outward, but those gravity waves are received in a reverse context by the opposed black hole. The gravitational radiation is quadrupolar, and the two-lobe shape is not unlike the Sun's magnetic field shape during the flip that takes place every 11 years, which is also known to be quadrupolar. The generation of the waveform in the strong field region is dominated by a "ringing region" similar to an acoustic mode or resonance in the Sun. This ringing region appears radially outside of the black-hole coalescence and appears during multiple stages of the dynamic evolution. Gravity pulses move both outward and inward at this ringing zone. If the ringing persists past the coalescence, and such ringing modes certainly exist in the Sun, it would help explain the Sun's structure and also be a factor in *hydrodynamics* – the mechanics of how the star both absorbs energy (or mass if one thinks of accretion) and radiates energy (and loses mass) into the outside world.

Asymptotic Flatness

Imagine a hyperbola. As it curves away from the center of the graph it gets closer and closer to the axis it travels along, but never touches the axis. Its limit along the axis is infinity. The same thing happens both outside and inside of a black hole. Outside, the gravity tapers off to zero, and the limit is "no gravity." Inside, gravity shoots up to infinity, and the limit is "infinite gravity."

A black hole therefore has two "asymptotically flat" ends. One approaches the horizontal axis traveling ever closer to zero but never quite getting there; and the other approaches the vertical axis, going infinite, but of course the maximum bit-resolution of even the most powerful computer will have failed long before that. In black hole simulations the infinite "asymptotically flat" ends can be cut or snipped to the event-horizon boundary – this is called "excision" and it prevents the equations from blowing up.

On the other end, far from the black hole, spacetime is flat and gravity goes to zero. The problem of determining initial values starts with flat spacetime, from the point of view of an outside observer. The parameters of mass, spin, and charge are observable from the outside and act as a reference for values determined inside the black-hole regime. Values of mass, spin, and charge are preserved in the black-hole framework of extreme curvature because they are well defined in the flat spacetime that surrounds the hole, to which the hole is "smoothly connected."

Mathematically, one can ask "how close is the mass of black-hole spacetime to the mass of flat spacetime?" That is, asymptotically, how close is the ratio of the two spacetimes to unity as x goes to infinity? Spatial infinity in the simulation is far away from the center of mass of the black-hole collision. As x goes to infinity, gravity goes to zero; and as x goes to zero, gravity goes to infinity. This means that far from the black hole, the mass of the black hole will approach the mass as seen by the outside observer. When both masses are the same, their ratio is 1, or unity. The way to say this is, "framework F is asymptotic to framework B" as x goes to infinity.

However, when x is small gravity approaches infinity, and this makes initial values of the simulation hard to determine. One way to solve the initial-value problem is to derive "estimates" using something called "Taylor's formula with remainder". The original function $f(x)$ must be differentiable. The estimate is a polynomial that approximates the differentiable function at any point a. The approximating polynomial is referred to as $f(a)$. Because Taylor's formula is iterative, using higher and higher-degree terms, the calculation stops at a certain point and the very small value left over, that would have been calculated with the remainder of the infinite series, is summed up in an approximating Taylor "remainder." The remainder, however, is not always easy to calculate.

By extending $f(x)$ from flat spacetime (framework F) into the black-hole regime (framework B) with approximations, one can correct the various parameters like mass, spin, and charge related to a particle's motion, that would otherwise be in error due to the coupling of the laws of motion to the geometry of spacetime. The result is a "global metric" which, in the ideal case, is valid all the way to the asymptotically flat "ends" near or inside the black-hole event horizon.

This idea of obtaining a global metric by asymptotically matching approximate solutions can be extended to sub-regions, by partitioning the space according to estimates of local curvature, and then matching the asymptotic Taylor expansions between regions. This may be the future of black-hole numerical computing. It's not clear to what degree Smarr used asymptotic estimates in his collision simulation, but there has been much research since then with some exciting results.

Thoughts

The registration of the two images (MDI and Smarr) was arrived at by matching contours and features: the simplest and most logical arrangement, where the extent of the simulation is also the extent of the Sun in the MDI image, has the most matches between features.

The structure at the center is fascinating, including resonant nodes or bands between the spindle and the tachocline in the MDI image; and in the simulation what look like inner rings around the very small central area, if the image is thought of as a cross-section. This would be the singularity, but knowledge about singularities besides their ring shapes is scanty. Unfortunately, there is almost no crossover between black-hole physics and solar physics. Lapse and shift has no counterpart in solar physics; just as lap time and shear find no counterpart in black-hole physics, except in a very specific mathematical sense.

The remarkable match between underlay and overlay is a testament to the integrity of Albert Einstein's general theory and his field equations. Eddington proved Einstein's theory when it came to the deflection of starlight around the Sun; the match between the black-hole theory (based on Einstein's equations) and the observational

MDI data is no less astounding. The underlay (left), being observed data, is perfectly matched by the overlay (right), which is entirely theory.

The question, as usual, is what is the Earth's role in the black-hole system? If the Van Allen belts are a black-hole ergosphere, then the singularity shown in the image would be the limits of Earth's place in the Sun. But the image doesn't give any hints about the existing dualism between the Earth and the Moon. Obviously if the Earth was going to be swallowed up by the black hole it would have been swallowed up 4.5 billion years ago – so that's not the concern. The image shows what things might look like from the outside; but all we know about the inside is what we observe from Earth, assuming that the Earth is inside of the Sun. Scientists know that the Van Allen belts are very much affected by solar disturbances. It would be interesting to know how our actions on Earth affect the Sun.

Some instruments have collected data on the solar interior but researchers have had a hard time drawing inferences from the data because it's so murky. And, to be honest, no doubt some of the data is classified: it would have to be. Hopefully both of these conditions will change, and let's hope sooner than later. With the Earth in so much environmental trouble, a new view of the Sun would be a wake-up call for our next generation of leaders.

The Sun can no longer be classified as a main sequence star, or even a main sequence star that has collapsed to a black hole. The Sun is probably a primordial black hole with Hawking charged-particle emission and radiation, which means it was born at the beginning of the universe. It's extremely unlikely that a regular binary star could have collapsed in perfect synchronization to produce the gravity signature of Smarr's controlled simulation. This implies that primordial black holes may start as binary black holes that collide and coalesce, or in some way may preserve the binary identity though appear from the outside as singletons. In either case, Smarr gave the world *not* a collapse but a collision. A collision differs from a collapse because there are no assumptions about the star's astrophysical history. The binary pair appears "ready made". The collision coalesces and results in a perfectly stable black hole in a singleton state. Smarr's simulation is the blueprint for a star that lives and breathes.

We must never take our little Earth for granted. Think of the complexity of everything above and below the thin skin of ground that we live on. No, we must respect our surroundings. James Van Allen, the researcher who mapped the belts around the Earth, remarked: "There is nothing that we as human beings can do that doesn't pale by comparison to the force of Nature." While that's true, it has a defeatist note to it that, with new discoveries in science, seems more and more out of place. It's not a very holistic attitude. A more attuned philosophy comes from David Spangler: "If there's one leap our consciousness has to make – and it's a quantum leap – it's how to be able to function, embody, look at and deal with multiple views of reality simultaneously and to be comfortable with that and not insist that reality only follow one particular course."

That's the challenge.

About the Author

Dan Browning lives in the San Francisco Bay Area and is possibly an amateur physicist but definitely a professional programmer. He enjoys thinking outside of the box and when not being a programmer enjoys hiking and beach combing along the California coast. He's also a rock hound / amateur geologist, and wishes he had discovered the Diavik and Ekati diamond mines in northern Canada.

Made in the USA
Monee, IL
06 May 2021

67943034R00062